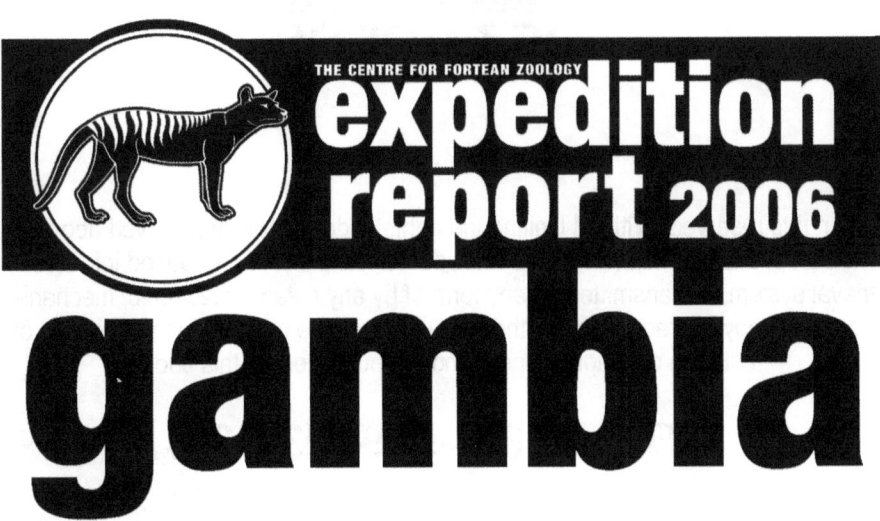

THE J.T. DOWNES MEMORIAL GAMBIA EXPEDITION 2006 - www.cfz.org.uk

THE CENTRE FOR FORTEAN ZOOLOGY
expedition report 2006
gambia

Edited by Corinna James and Jonathan Downes
Cover and internal design by Mark North for CFZ Communications
Using Microsoft Word 2000, Microsoft , Publisher 2000, Adobe Photoshop CS.
Cover photograph by Suzi Marsh

First published in Great Britain by CFZ Press

CFZ Press
Myrtle Cottage
Woolfardisworthy
Bideford
North Devon
EX39 5QR

CFZ PRESS

© CFZ MMVI

All rights reserved. Without limiting the rights under copyright reserved above, no part of this publication may be reproduced, stored in or introduced into a retrieval system, or transmitted, in any form of by any means (electronic, mechanical, photocopying, recording or otherwise), without the prior written permission of both the copyright owners and the publishers of this book.

ISBN: 978-1-905723-03-4

"There is always something new out of Africa."

Pliny the Elder, (23 AD - 79 AD)

CONTENTS

INTRODUCTION by Jonathan Downes	7
EDITOR'S NOTES	9
Foreword by Dr. Karl P.N. Shuker	11
Richard Freeman	21
Chris Moiser	67
Lisa Dowley	117
Oll Lewis	165
Suzi Marsh	201
Chris Clark	223
APPENDIX ONE: CRESTED SERPENTS WORLDWIDE by Richard Freeman	239
APPENDIX TWO: FRESHWATER ECOLOGY OF THE GAMBIA by Simon Wolstencroft	267
Acknowledgments and Picture Credits	275

INTRODUCTION

by Jonathan Downes
(Director, Centre for Fortean Zoology)

The Centre for Fortean Zoology is the only professional, scientific and full-time organisation in the world dedicated to Fortean Zoology; a portmanteau discipline which includes cryptozoology - the study of unknown animals. Since 1992, when it was founded, the CFZ has carried out an unparalleled programme of research and investigation all over the world. Because we are funded purely from private subscription, we feel that it is only appropriate that we make all our findings and research public as well. This is the first, of what we hope will be a long line, of CFZ expedition reports published in book form, and as Director of the CFZ it gives me great pleasure to write this introduction.

I first heard about the mystery of the beast of Bungalow Beach in the early 1990s, about the time that I first shocked my family and friends by talking about my plans for starting up a worldwide society for those interested in cryptozoology. In 1994 I made friends with Dr Karl Shuker, and we met for the first time. He told me more about the incident, and a few years later, introduced me to Owen Burnham.

I was determined then, that one day, the CFZ would go to The Gambia and dig up the beach. I didn't realise that it would take over a decade before we managed it, but the best laid plans of Animals & Men do tend to go pear shaped on occasion.

However, this project seemed to me to be an ideal one for the CFZ. Whilst it would be wonderful to have the resources to mount a major expedition in search of the yeti - for example - as Heuvelmans himself pointed out over half a century ago, the ogistical problems inherent in doing the thing properly would cost a small fortune. As the CFZ expands, we are getting closer to the day when major expeditions with a five or six figure budget will be a reality, but in the meantime, smaller, more focussed expeditions like this one, are the sort of projects we ought to be pursuing.

Over the years we collected reams of additional data on the cryptozoology of The Gambia, and when we found out that as well as the beast of Bungalow Beach, there were also the Ninki Nakna, and Armitage's Skink to be investigated, it was proposed at the annual planningmeeting in January 2006 that this should be the major expedition of the year. The proposal was carried unanimously, and I don't think that anyone has had any regrets since.

As you will see, upon reading this book, the expedition was a success.

But the 2006 Gambia expedition is important to me on a personal level as well.

My father, John Tweddell Downes, I.S.O, died on the 14th February 2006. He had

become a big supporter of the CFZ, and during his final illness he liked nothing more than to have my friend and colleague Richard Freeman and me sitting on his bed, telling stories of our adventures, and plotting our plans for the future. He spent many years living in West Africa, and was particularly interested in - what was then - the forthcoming expedition to The Gambia. He had been planning to donate some money towards it, but died before he was able to do so, and so my brother and I decided that some of the money donated at his funeral service should go towards the expedition, which we decided to sponsor in his name...

For those of you interested, here is an excerpt from his obituary:

"John Downes was born in Plymouth in 1925. He joined the Merchant Navy in 1943 and served as a Communication's Officer during the Battle of the Atlantic. He returned to shore in 1947, and after marrying his childhood sweetheart Mary, worked for the Ministry of Agriculture in North Devon. In 1952 he joined the Colonial Service in Nigeria and together with Mary, worked in some of the most isolated parts of North Nigeria on the southern borders of the Sahara Desert. Often they were the first Europeans to have visited these remote regions for over half a century.

In 1960 John was transferred to Hong Kong where he rose to the rank of Assistant Colonial Secretary, and as a commisioned officer in the Royal Naval Reserve, founded the Hong Kong Sea Cadet Corps. His outstanding work within the Civil Service was duly recognised when he was made a companion to the Queen and awarded the Imperial Service Order. Unfortunately John was forced to take early retirement on medical grounds in 1971, and he and Mary spent the rest of their lives together in Woolfardisworthy, North Devon, where they quickly earned the respect and love of the local community. John became a financial manager for many local businesses, and was a tireless pillar of the community and church. He renewed his love of the sea by becoming the Commanding Officer of *TS Revenge*, the Bideford Sea Cadet Corps and became an inspiration for generations of young people.

During his retirement, John became an acknowledged author and was an expert in such diverse areas as African History, and Devonshire dialect.

He was the author of several books including `A Dictionary of Devonshire Dialect` (1988), `Granfer's Bible Stories` (2005) and `Fragrant Harbours, Distant Rivers` (2006). He was widowed in 2002, and after a long battle with Parkinson's Disease, died peacefully in North Devon District Hospital on Tuesday February 14th aged 81. He is survived by his two sons, Jonathan Downes, (46) - Director of the Centre for Fortean Zoology - and The Rev'd Richard Downes BEM CF (42), a Chaplain to the Forces in the Army. "

It is, therefore, with great pleasure that I dedicate this book to the memory of a great man.

EDITOR'S NOTES

This current volume was prepared in a great hurry, in order to have it available for the first public appearances by the team upon their return to the UK. However, because of this unwarranted haste, several corners had to be cut. We decided that rather than editing all the different accounts into a single narrative, it was better to provide all five accounts separately. This we have done, and whilst every effort has been made to standardise all spellings and grammar, it is almost certain that some anomalies will remain. However, it makes for interesting reading. Just as in *Rashomon* [1] - a 1950 Japanese motion picture directed by Akira Kurosawa (in collaboration with Kazuo Miyagawa) and starring Toshiro Mifune.- the same story being told from a variety of different perspectives, allows us a uniquely fortean view of events. Some accounts are matter-of-fact and scientific, others more subjective, but they add up to a fascinating whole.

We have included the taxonomic name of each animal species mentioned in the text, but only on its first mention. All subsequent entries are left purely in English..

---

1. Based on two stories by Ryūnosuke Akutagawa , it describes a rape and murder through the widely differing accounts of four witnesses, including the perpetrator and, through a medium, the murder victim. The story unfolds in flashback as the four characters—the bandit Tajōmaru (Mifune), the murdered samurai Kanazawa-no-Takehiro (Masayuki Mori), his wife Masago (Machiko Kyō), and the nameless Woodcutter (Takashi Shimura)—recount the events of one afternoon in a grove. But it is also a flashback within a flashback, where the woodcutter or priest has told what each individual said at the court. Each story is mutually contradictory, leaving the viewer unable to determine the truth of the events, and should be required viewing for every fortean.

FOREWORD

by Dr. Karl P.N. Shuker

Doesn't time fly when you're having fun! As I write this foreword, I can scarcely believe that 20 years have gone by since I penned what became my very first investigative cryptozoological article, published as a two-parter in the September and October 1986 issues of a now long-defunct British magazine called *The Unknown*. And what was my article's subject? Why, none other than a certain mysterious sea beast found dead a few years earlier on a beach in the Gambia - the very same creature that I am writing about now in this foreword, as it was one of the major quarries of this important new cryptozoological expedition launched by the Centre for Fortean Zoology (CFZ). Clearly, time not only flies but also on occasion takes delight in looping the loop!

Back in 1986, I became the first cryptozoologist to write about the Gambian sea serpent, and went on to document it further in a number of other publications, including various of my books - most extensively of all within *In Search of Prehistoric Survivors* (1995). Indeed, this remarkable case launched my career as a writer, and although I have since introduced a sizeable number of other hitherto little-publicised or wholly unpublicised cryptids to the general international reading public - such as the Sri Lankan horned jackal and devil bird, Goodenough Island mystery bird, New Guinea devil pig, Scottish earth hound, Indonesian veo and horned cat, New Caledonian du, Irish dobhar-chú, Shatt al Arab venomous mystery fish, Zanzibar makalala, Ethiopian death bird, Zululand kondlo, Arctic North American waheela, Kellas cat, Mongolian death worm, Hungarian reedwolf, Fujian blue tiger, Welsh cenaprugwirion, bigfin mystery squid, St Helena sirenian, Timor Sea ground shark, and a certain coxcombed serpent (see later), among others - `Gambo` (as it was subsequently dubbed, though not by me, I hasten to add!) remains one of the most intriguing, tantalising, and controversial cryptids that I have ever investigated.

Consequently, it seemed appropriate to include within this foreword the history and facts of the Gambo case, thereby setting the scene for the recent search conducted by the CFZ's 'J.T. Downes Memorial West Africa Expedition 2006', documented within the main body of this volume. And what better way to do this than to reproduce my own account of it that appeared in my book *In Search of Prehistoric Survivors*, because this incorporates an extensive first-hand description given to me by none other than Owen Burnham, Gambo's principal eyewitness:

On 12 June 1983, wildlife enthusiast Owen Burnham and three family members encountered the carcase of a huge sea creature, washed up onto a beach in Gambia, western Africa. Most sea monster remains are discovered in an advanced state of decomposition, greatly distorting their appearance and making positive identification very difficult, but the carcase found by Burnham was exceptional, as it was largely intact, with no external decomposition.

Now residing in England but having lived most of his life in Senegal, Burnham was very familiar with all of that region's major land and sea creatures, but he had never seen anything like this before. Realising its potential zoological significance, he made meticulous sketches and observations of its outward morphology, and noted all of its principal measurements.

In May 1986, a British magazine published a short account by Mr Burnham describing his discovery, and including versions of his original sketches. Greatly interested, I wrote to him, requesting further details, in order to attempt to identify this remarkable creature. During our correspondence, Burnham kindly gave me a comprehensive description (plus his sketches) of its appearance. The following is an edited transcript of Burnham's first-hand account of his discovery, prepared from his letters to me of May, June, and July 1986:

> I grew up in Senegal (West Africa) and am an honorary member of the Mandinka tribe. I speak the language fluently and this greatly helped me in getting around. I'm very interested in all forms of life and make copious observations on anything unusual.
>
> In the neighbouring country of Gambia we often went on holiday and it was on one such event that I found this remarkable animal.
>
> June 1983. An enormous animal was washed up on the beach during the night and this morning [June 12] at 8.30 am I, my brother and sister and father discovered two Africans trying to sever its head so as to sell the skull to tourists. The site of the discovery was on the beach below Bungalow Beach Hotel. The only river of any significance in the area is the Gambia river. We measured the animal by first drawing a line in the sand alongside the creature then measuring with a tape measure. The flippers and head were measured individually and I counted the teeth. [In the sketches accompanying his description, Burnham provided the following measurements: Total Length = 15-16 ft; Head+Body Length = 10 ft; Tail Length = 4½-5 ft; Snout Length = 1½ ft; Flipper Length = 1½ ft.]
>
> The creature was brown above and white below (to midway down the tail).
>
> The jaws were long and thin with eighty teeth evenly distributed. They were similar in shape to a barracuda's but whiter and thicker (also very sharp). All the teeth were uniform. The animal's jaws were very tightly closed and it was a job to prise them apart.
>
> The jaws were longer than a dolphin's. There was no sign of any blowhole but there were what appeared to be two nostrils at the end of the snout. The creature can't have been dead for long because its eyes were clearly visible and brown although I don't know if this was due to death. (They weren't protruding). The forehead was domed though not excessively. (No ears).
>
> The animal was foul smelling but not falling apart. I've seen dolphins in a similar state after five days (after death) so I estimate it had been dead that long.
>
> The skin surface was smooth, the only area of damage was where one of the flippers (hind) had been ripped off. A large piece of skin was loose. There were no mammary glands present and any male organs were too damaged to be recognizable. The other flipper (hind) was damaged but not too badly. I couldn't see any bones.
>
> I must mention clearly that the animal wasn't falling apart and the *only* damage was in the area (above) I just mentioned. The only organs I saw were some intestines from the damaged area.

The paddles were round and solid. There were no toes, claws or nails. The body of the creature was distended by gas so I would imagine it to be more streamlined in life. It wasn't noticeably flattened. The tail was rounded [in cross-section], not quite triangular.

I didn't (unfortunately) have a camera with me at the time so I made the most detailed observations I could. It was a real shock. I couldn't believe this creature was laying in front of me. I didn't have a chance to collect the head because some Africans came and took the head (to keep skull) to sell to tourists at an exorbitant price. I almost bought it but didn't know how I'd get it to England. The vertebrae were very thick and the flesh dark red (like beef). It took the men twenty minutes of hacking with a machete to sever it.

I asked the men on the scene what the name of this animal was. They were from a fishing community and gave me the Mandinka name *kunthum belein*. I asked around in many villages along the coast, notably Kap Skirring in Senegal where I once saw a dolphin's head for sale. The name means 'cutting jaws' and is the term for dolphin everywhere. Although I gave good descriptions to native fishermen they said they had never seen it. The name *kunthum belein* always gave [elicited] a dolphin for reply and drawings they made were clearly that. I also asked at Kouniara, a fishing village further up the Casamance river but with no success. I can only assume that the butchers called it by that name due to its superficial similarities. In Mandinka, similar or unknown animals are given the name of a well known one. For example a serval is called a `little leopard`. So it obviously wasn't common. I've been on the coast many times and have never seen anything like it again.

I wrote to various authorities. [One] said it was probably a dolphin whose flukes had worn off in the water. This doesn't explain the *long pointed tail* or lack of dorsal fin (or damage).

[Another] decided it could be the rare *Tasmacetus shepherdi* whose tail flukes had worn off. This man mentioned that the blow hole could have closed after death. Again the tail and narrow jaws seem to conflict with this. *Tasmacetus*'s jaws aren't too long and the head itself seems to be smaller than my animal's. *Tasmacetus* has two fore flippers and none in the pelvic region. The two flippers are quite small in relation to body size and pointed rather than round. *Tasmacetus* has a dorsal fin and 'my' animal didn't seem to have one or any signs of one having once been there. *Tasmacetus* even without tail flukes wouldn't have a tail long enough or pointed enough. The tail of the animal I saw was very long. It had a definite point and didn't look suited for a pair of flukes. Apparently, *Tasmacetus* is brown above and white below and this seems to be the only link between the two animals. I've been to many remote and also popular fishing areas in Senegal and I have seen the decomposing remains of sharks and also dead dolphins and this was so different.

[A third] said it must have been a manatee. I've seen them and believe me it wasn't that. The skin thickness was the same but the resemblance ended there.

Other authorities have suggested crocodiles and such things but as you see from the description it just can't have been.

After I think of the coelacanth I don't like to think what could be at the bottom of the sea. What about the shark (*Megachasma*) [megamouth shark] which was fished up on an anchor in 1976?

I looked through encyclopaedias and every book I could lay hands on and eventually I found a photo of the skull of *Kronosaurus queenslandicus* which is the nearest thing so far. Unfortunately the skull of that beast is apparently ten feet long and clearly not of my find.

The skeleton of *Ichthyosaurus* (not head) is quite similar if you imagine the fleshed animal with a pointed tail instead of flukes. I spend hours at the Natural History Museum [in London, England] looking at their small plesiosaurs, many of which are similar.

I'm not looking to find a prehistoric animal, only to try and identify what was the strangest thing I'll ever see. Even now I can remember every minute detail of it. To see such a thing was awesome.

Presented with such an amount of morphological detail, quite a few identities can be examined and discounted straight away - beginning with *Tasmacetus shepherdi*. Although somewhat dolphin-like in shape, this is a primitive species of beaked whale, described by science as recently as 1937, and known from only a handful of specimens, mainly recorded in New Zealand and Australian waters, but also reported from South Africa. Whereas all other beaked whales possess no more than four teeth (some only have two), *Tasmacetus* has 80, and its jaws are fairly long and slender.

However, the Gambian beast's two pairs of well-developed limbs effectively rule out all modern-day cetaceans as plausible contenders, because these species lack hind limbs. They also eliminate the archaeocetes - even *Ambulocetus*. For although this 'walking whale' did have two well-formed pairs of limbs, unlike the Gambian sea serpent its teeth were only half as many in number, yet of more than one type. Its long tail and dentition effectively ruled out pinnipeds and sirenians from contention too.

Many 'sea monster' carcases have proven, upon close inspection, to be nothing more exciting than badly-decomposed sharks, but as the Gambian beast displayed no notable degree of external decomposition this 'pseudoplesiosaur' identity was another non-starter.

Indeed, after studying his detailed letters and sketches, it became clear that, incredibly, the only beasts bearing any close similarity to Burnham's Gambian sea serpent were two groups of marine reptilians that officially became extinct over 60 million years ago.

One of these comprised the pliosaurs [short-necked long-headed plesiosaurs] - thus including among their number the mighty Australian *Kronosaurus* that Burnham himself had mentioned - but whereas their nostrils' external openings had migrated back to a position just in front of their eyes, those of the Gambian sea serpent were still at the tip of its snout

The other constituted the thalattosuchians [prehistoric marine crocodilians] - always in contention here, on account of their slender, non-scaly bodies, paddle-like limbs, and terminally-sited external nostrils. True, their tails possessed a dorsal fin, but a thalattosuchian whose fin had somehow been torn off or scuffed away would bear an amazingly close resemblance to the beast depicted in Burnham's sketches. Alternatively, assuming that a thalattosuchian lineage had indeed persisted into the present day, its members may no longer possess such a fin anyway.

Without any physical remains of the beast available for direct examination, however, its identity can never be categorically confirmed.

Little did I realise, when writing that account back in 1994, that 12 years later a serious cryptozoological expedition would be returning to the very spot where Gambo's

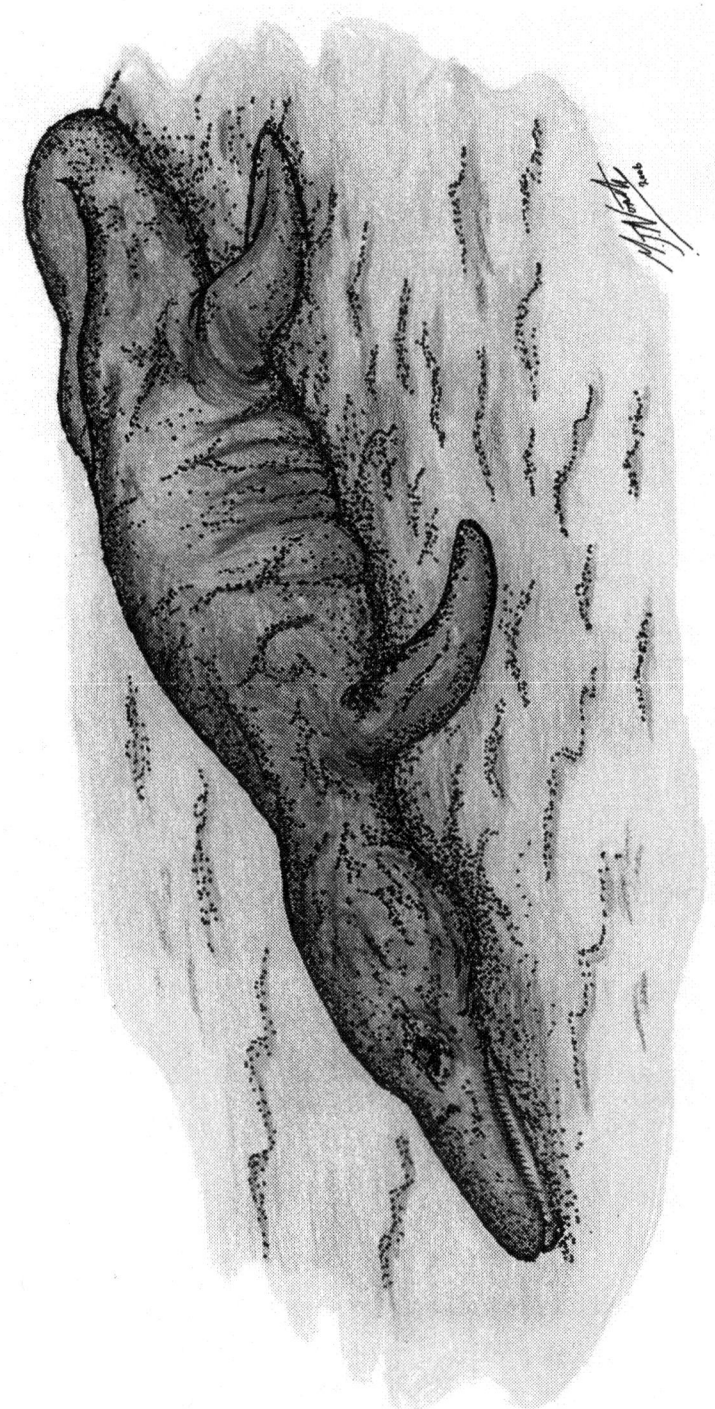

Reconstruction after Owen Burnham
by Mark North

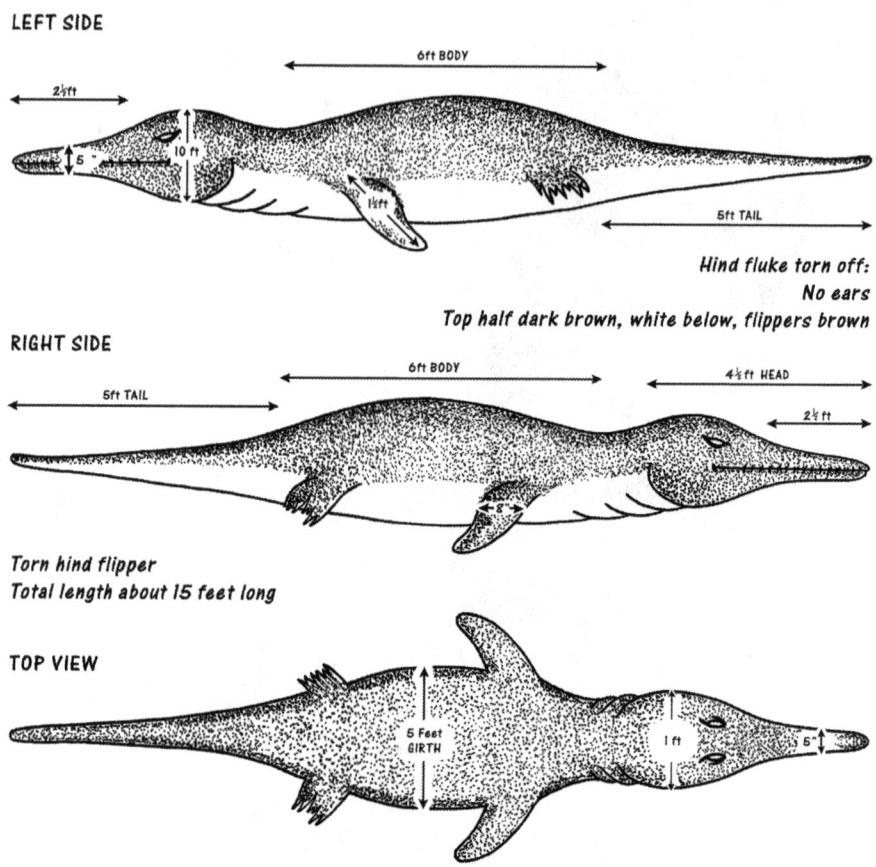

remains had been buried in the first-ever attempt to uncover and conclusively identify this enigmatic creature. And did they succeed? You'll have to read their report and find out for yourself!

As for myself, 20 years on from my first article I remain totally open-minded as to what Gambo was. Contrary to a number of claims or assumptions made by others over the years, I have never stated that I believe it to have been a modern-day descendant of a prehistoric lineage of reptile. I have merely stated that, based upon Burnham's verbal description and sketches, that is what it most closely resembles. But as the saying goes, appearances can (and often do) deceive, so it would be ridiculous to make any firm assertion as to its identity without having first examined physical evidence - which is why I have never done so, and have no intention of ever doing so.

After all, it is possible (though in my view unlikely) that Burnham's description and drawings are not very accurate, in which case Gambo may have been nothing more after all than some ordinary, known species of cetacean; or, at most, a previously unknown cetacean species - in which case I propose *Gambiocetus burnhami* gen. nov. sp. nov. ('Burnham's Gambian whale') as a suitable scientific name for it. In any event, one record firmly set straight, I trust!

Making the opportunity to write this foreword doubly pleasurable is that the expedition also uncovered information, this time concerning a West African mystery reptile known as the ninki nanka, that has relevance to another longstanding cryptozoological favourite of mine - the crowing crested cobra. This extraordinary cryptid is said to resemble an extremely large snake but adorned with a coxcomb and facial wattles like those of a farmyard rooster and further likened to the latter fowl by its ability to crow like a rooster too. It has been reported widely across tropical Africa, but chiefly in the east and south, which makes the CFZ expedition's findings in West Africa particularly interesting and significant.

Thanks in no small way to the internet, there are now more armchair cryptozoologists than ever before, discussing across cyberspace all manner of cryptozoological conundra, which should be a very positive, informative experience.

Sometimes, however, such exchanges can degenerate into scenarios whereby individuals become embroiled in ludicrously heated exchanges as to what a given cryptid can and cannot possibly be, yet usually without ever having made the slightest effort to go out into the big world beyond their computer screens and actually seek the said beastie in the field for themselves, or even interview a few eyewitnesses who have encountered it. This is why I am so pleased that there are also real cryptozoologists out there, like those who took part in this expedition, who take the time and effort to do just that, and, in so doing, add enormously to our knowledge and understanding of this planet's unrevealed fauna.

As you will discover for yourselves, this expedition's reports make fascinating reading, containing much new data never before made available, and show how cryptozoology should be conducted.

The expedition was named in honour of the late father of CFZ Director and Founder Jonathan Downes, and I have no doubt that he would have been very proud of and excited by its success. Congratulations to everyone involved, and I am already eagerly awaiting your return expedition to West Africa in 2007. Carry on cryptozoologising!

Dr. Karl Shuker

The J.T. Downes Memorial Gambia Expedition 2006

Team

Back row - Suzi Marsh, Chris Moiser, Oll Lewis
Front row - Chris Clark, Lisa Dowley, Richard Freeman

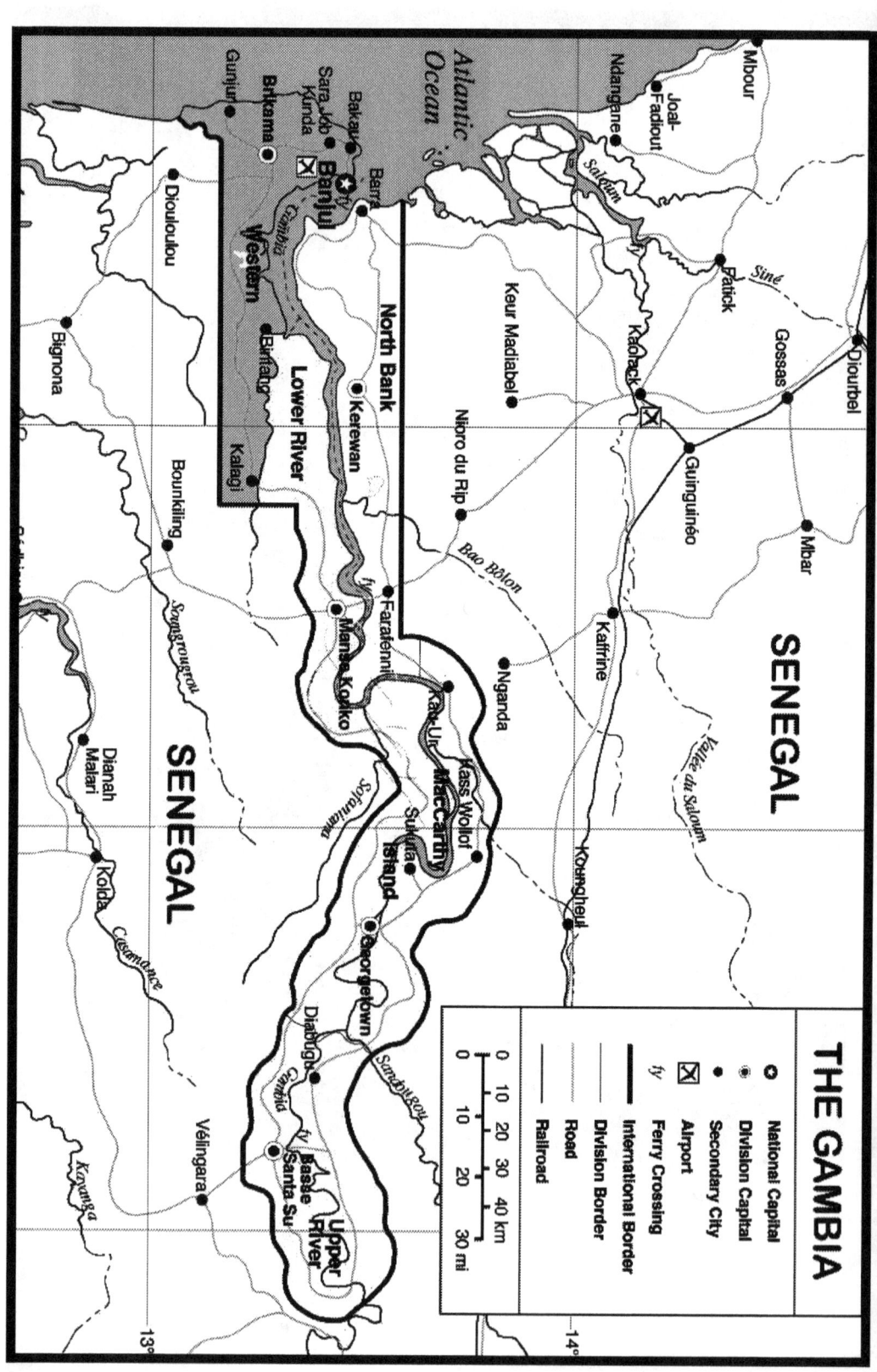

Richard Freeman

Richard Freeman is one of Britain's few professional cryptozoologists. His interest in unknown animals reaches back to his childhood and he has had a long and varied career working with exotic animals. After leaving school he worked as a zookeeper at Twycross Zoo in the Midlands where he became head curator of reptiles. During this time he bred many rare and endangered species. After leaving the zoo in 1990 he worked in several specialist pet shops and an animal sanctuary. All in all he has worked with over 400 species of animal.

He has travelled extensively in East Africa, Europe and the US studying the native fauna. In 1996 he took a degree in zoology at Leeds University and afterwards, moved to Exeter to work full time at the Centre for Fortean Zoology.

In October 2000 he visited the remote jungles, rivers, and caves of Northern Thailand in search of the Naga, a giant crested serpent said to lurk in the primal morasses of Indo-China, and explored deep caves never visited by Westerners before. He subsequently carried out expeditions to Sumatra (2003 and 2004), in search of the bipedal ape orang pendek, and Mongolia (2005), in search of the semi-legendary deathworm.

His interests include Doctor Who and other British television science fiction, collecting Batman comics, Forteana and gothic rock music. He is the author of *Dragons: More than a myth?* (2005) and *Explore Dragons* (2006).

Day 1, 4th July

As the plane descended after a flight, during which I had spent most of my time asleep, a view of extensive mangrove swamps appeared. After the dull greys and browns of the Sahara, the green was striking. The mangroves are thick, but are they thick enough to hide a dragon?

Soon the mangroves gave way into cultivated land and buildings. There seemed little in the way of transition from one to the other. This is very different from Sumatra, where twixt jungle and farm land there is the semi-wild 'garden'.

Wildlife seems to have no problem adapting to urban life. Hooded vultures and black kites crowded the skies. Huge swallow-tailed butterflies flitted past as we disembarked.

It feels strange not to be in Asia. All of my major expeditions have been to the Orient. It is 21 years since I was in Africa. The `Dark Continent` is the home of travellers' tales and monsters but is there any truth to them?

The capital, Banjul, has the look of the more down-at-heel parts of Pedang [1]. The African Village Hotel is certainly not the Hilton, and the dim lights, the termite colony in the toilet, and tatty mosquito net would not have earned it a single star in Britain, but compared to some of the hotels in Indonesia and the Gobi, it is a positive relief. It is also preferable to sleeping on the jungle floor, or on desert sands.

A short walk to the beach rewarded us with a look at some magnificent rainbow agamas (*Agama agama*) and five-lined skinks *(Euprepis margaretifera)*. Tomorrow we test the waters with a view to excavating the carcass of `Gambo`, if it still exists.

Day 2, 5th July

After breakfast we walked down to Bungalow Beach where, in 1983, Owen Burnham was said to have discovered the carcass of a strange marine creature.

On the way to the beach a local man, who called himself 'Gabby', began to follow us and engage us in conversation. Though totally uninvited, he followed us down the beach as an un-asked for guide.

Unfortunately, in the intervening 23 years, Bungalow Beach has become a lot less remote. Several hotels have sprung up, and also a nightclub called *Destiny's*. The nightclub's beer garden looks as if it may well have been erected right over `Gambo`'s resting place! As the club belongs to the president's brother, [1] who is being investigated for fraud, any request that we hack through the concrete with

---

1. Padang is the capital and largest city of West Sumatra, Indonesia. It is located on the western coast of Sumatra at 0°57′ S 100°21′E. It has an area of 694.96 km² and a population of over 750,000 people, mostly speakers of the Minangkabau language. Richard and Chris Clark stayed there during the Sumatran expeditions of 2003 and 2004.

picks might not go down too well.

It would seem that `Gambo` has become the Schrödinger's cat [2] of cryptozoology, existing either just under the foundations of the beer garden, or just outside of it.

We met some stall-holders, including a wood carver. On enquiring about Ninki Nanka carvings, a one of them said he could have one ready in four days, for the extortionate price of 800 Dalasi (about £16). He showed me a photograph of one of these Ninki Nanka carvings, but it was just a generic Chinese dragon, rather than something from local folklore, so I declined.

A little later on we came to another market at Kotu. Here jewellers have been making silver Ninki Nankas for years. Chris Moiser knew one of them.

One of the jewellers, Baka Samba, or Mr Fixit as he is locally known, was a mine of information. He had not seen Ninki Nanka, but his late uncle, a renowned hunter, had. Many years ago his uncle had seen the dragon up-river. He died within five years of the sighting, and Baka was convinced that seeing the dragon had sealed his fate. The description his uncle gave was vague. *"Huge and terrible"* was one phrase. It possessed four legs and a horrible head, and he said that it had fire in its mouth. This could relate to the flickering of a brightly coloured tongue, or even a brightly coloured interior of the mouth. Perhaps the death was brought on by malaria, contracted in the remote region where the sighting occurred?

Baka's nephew, Baba, told us that Ninki Nanka resembled a crocodile, but its head

---

1. Yahya (Abdul-Aziz Jemus Junkung) Jammeh (born May 25, 1965) is the President of The Gambia. As chairman of the Armed Forces Provisional Ruling Council, he took control of the country in a military coup in July 1994, and was elected as president two years later, in September 1996, in widely criticized elections. He founded the Alliance for Patriotic Reorientation and Construction as his political party. Jammeh was re-elected on October 18, 2001 with about 53% of the vote; this election was generally deemed free and fair by observers.

President Jammeh was formerly named **Yahya Alphonse Jemus Jebulai Jammeh**.
Jammeh has been accused of restricting freedom of the press; harsh new press laws were followed by the unsolved killing of a reporter who had been critical of them, Deyda Hydara, in December 2004.

2. Schrödinger's cat is a seemingly paradoxical thought experiment devised by Erwin Schrödinger that attempts to illustrate the incompleteness of an early interpretation of quantum mechanics when going from subatomic to macroscopic systems. The experiment proposes:

A cat is placed in a sealed box. Attached to the box is an apparatus containing a radioactive atomic nucleus and a canister of poison gas. This apparatus is separated from the cat in such a way that the cat can in no way interfere with it. The experiment is set up so that there is exactly a 50% chance of the nucleus decaying in one hour. If the nucleus decays, it will emit a particle that triggers the apparatus, which opens the canister and kills the cat. If the nucleus does not decay, then the cat remains alive. According to quantum mechanics, the unobserved nucleus is described as a superposition (meaning it exists partly as each simultaneously) of "decayed nucleus" and "undecayed nucleus". However, when the box is opened the experimenter sees only a "decayed nucleus/dead cat" or an "undecayed nucleus/living cat."

The question is: when does the system stop existing as a mixture of states and become one or the other? The purpose of the experiment is to illustrate a paradox; as Schrödinger wrote,

"The [wavefunction] for the entire system would [have] the living and the dead cat (pardon the expression) [sic] mixed or smeared out in equal parts."

and teeth were shaped differently and it had big eyes.

I asked Baka if his shop had been here in 1983. He confirmed that it had, and I went on to ask him if he recalled a large creature having been washed up on the beach in '83. He told me that he did remember this, and that the creature was a large dolphin. He indicated its length by pointing to a breeze-block, and saying it was as long as "from here to there". This was about 9 feet. He told us that it was still alive when it had been washed up, and that some white men in a boat had tried to rescue it. However, it vomited and died on the beach, whereupon it was buried.

Oll produced a note-book and we asked Baka to draw it, but he said he could not draw. Therefore Oll drew as Baka directed. He produced a sketch of a dolphin *sans* the dorsal fin. Baka showed us a silver dolphin, and indicated that the creature was like this, minus the dorsal fin. Could this have been `Gambo`?

Later, whilst drinking fruit-juice in the Bungalow Beach Hotel, Chris Clark and I were approached by a man called Papa. He claimed to know of a lake in Senegal that was inhabited by a Ninki Nanka. He had seen strange lights in it by night and weird disturbances by day. It was 250 km away, but he said we could hire a car to go there.

The lake lay between the villages of Ganatit and Kanut in an area known as Bangagara. He also claimed to have seen the dragon's tracks around the lake. He drew a copy in the sand. They looked like the drag-marks of a crocodile, and he indicated that they were about four feet wide. Chris asked him if the lake had a bad smell, and contained a lot of rotting vegetation. Papa answered yes to both questions. Chris thought that a build up of methane from rotting plant matter could be the cause of the strange occurrences.

Papa could not give a price on how much it would cost to get there, but gave Chris his number. We all agreed that it was a little too far out of the way, and would probably be too expensive.

Back at the hotel we met a medical photographer called Ian Clifford who had a special interest in snakes. He was in the Gambia to photograph venomous snakes.

Tomorrow we are to visit Abuko National Park where Ninki Nanka allegedly killed a night watchman in 1912, and was subsequently driven away by the sight of its own reflection in a huge mirror erected on the advice of a witch-doctor.

Day 3, 6th July

We travelled to the park with Chris Moiser's taxi driver friend, Assan. Abuko is Africa's smallest National Park at 105 hectares. It is, however, teeming with wildlife. I enquired of our guide Musa Jatta if he knew of Ninki Nanka. He said that Ninki Nanka was like a huge python, big enough to swallow a whole cow. It had legs and wings like a bat's wings, and could breathe fire, and could move around on land and in the water, but he did not know if it could fly. Musa went on to tell me that, sometime between 2001 and 2003, a dragon had caused a lorry to crash when it had crossed a road, leaving a great furrow in its wake. It was during heavy rain that the creature was on the move.

Abuko is an excellent reserve, swarming with wildlife. The pools were low due to the dry season having just ended, but the rainy season having not yet kicked in. They did not look big enough to hide an animal the size of a Ninki Nanka. However, they were inhabited by Nile crocodiles *(Crocodylus niloticus)* and Nile monitors, *(Varanus niloticus)* with African jacanas, *(Actophilornis africanus)*, hamerkop stork, *(Scopus umbretta)* pied kingfisher *(Ceryle rudis)* and malachite kingfisher, *(Alcedo cristata)*, pigmy kingfisher, *(Ceyx Picta)* blue *cordon bleu* finch, *(Uraeginthus cyanocephala)*, lilac roller, *(Coracias caudate)* and green touraco *(Tauraco persa)* all putting in an appearance. Red colobus monkeys *(Piliocolobus badius)* crashed through the branches.

As we moved on, I turned a corner in a jungle path and saw a snake. As it reared up, I realised that it was a cobra. [1] Musa yanked me back yelling *"It's a spitting cobra, it can kill you man!"* but I was far too excited to be scared. It spat at us once, turned away briefly, turned back, and spat again, then finally slithered into the undergrowth. Sadly, it all happened so fast, that I did not get the cobra on film.

Though not as tall or thick as true rainforest, the jungle at Abuko (dry forest) is still spectacular. Tall palms and huge wild mangos, encircled by strangler vines, were interspersed with massive termite mounds.

Chris Moiser had made some enquiries about the story of the night watchman, and turned up an amazing fact; his grandson worked as a guide at Abuko!

We found the man in question - a tall, amiable chap in his mid-20's called Hassan Jinda. Hassan related his grandfather's strange story, which – not unusually for the results of CFZ fieldwork, in our experience – showed that much of that had been written on the subject in the west, was wrong. In this instance it was the date:

---

1. Spitting cobra refers to any one of several species of cobras that have the ability to spit or eject venom from their mouth when defending themselves against predators. The spit venom, although not generally fatal on contact, can cause permanent blindness if introduced to the eye (causing chemosis and corneal swelling[1]), and/or skin scarring if left untreated.

Despite their name, these snakes don't actually spit their venom. They rather spray the venom, using muscular contractions upon the venom glands. The muscles squeeze the glands to forcefully propel the venom out of front openings in the fangs. To further aid this, the two streams of venom from each fang cross each other soon after leaving the snake, adding velocity to the combined stream. When cornered, some species can "spit" their venom up to a distance of two meters. Four out of seven species of cobras found in Africa and seven out of nine species found in Asia can spit to varying degrees.

While spitting is typically their primary form of defence, all spitting cobras are also capable of delivering venom through a bite as well. Most species' venom exhibit significant hemotoxic effects, along with more typical neurotoxic effects of other cobra species.

The date we had from the *Lonely Planet* guide was wrong. Hassan's granddad, Papa Jinda, saw the dragon twice - firstly in 1943 then again in 1947.

Back in the 1940's, the area that is now Abuko was used to provide water to Bathurst (now Banjul) [1]. One night Papa saw that something had damaged the water pipes at the pumping station. Then he saw the culprit, a Ninki Nanka. It was unimaginably huge, and when its head was in sight its tail could not be seen. Its body was covered with scales that shone like diamonds and its head bore a crest of fire. As it moved it would pause, look up and down, and then move on as if searching for something.

Papa thought it was a female dragon, as the crests of male and female Ninki Nanka were different. The male's crest looks like a crown, the female's like a tiara. According to local belief, if you see a male dragon you die immediately, but if you see a female it can take you up to five years to die.

Papa saw the dragon again in 1947. After this encounter he had pains in his legs and sides, and his hair began to fall out. He died two weeks later.

Hassan said that he knew a man who had what he thought was Ninki Nanka scales. The old man had found them in Abuko several years before. We were taken to the man who said he had discovered the scales on the jungle floor. He kept them in a pouch about his neck and used them to make charms. He allowed us to look at them and we could see that they were transparently not true scales and looked man-made in origin. To be thorough, we bought a scale from the old man to bring back to England for analysis. It was a silvery, irregular, translucent flake, about an inch across and it was the general consensus that it had been a piece of decayed celluloid. The man said that it had been red and blue in colour when he had found it.

Another one of the guides said that he knew where a Ninki Nanka is supposed to live to this very day. This was in an area of swamps about 90 km away, which seemed a better bet than the lake in far Senegal.

Day 4, 7th July

After a day of jungle trekking, we decided to have an easier day. The best-laid plans of mice and men…

Lisa, Chris Clark, and I thought we would have a look at the local fish-market. I had made a drawing based on the original `Gambo` sketch by Owen Burnham to see if the fishermen recognised it.

---

1. **Banjul** is the capital of The Gambia. The population of the city proper is only 34,828 but the total urban area is many times larger with a population of 523,589 (2003 census). It is located on St Mary's Island (or *Banjul Island*) where the Gambia River enters the Atlantic Ocean. Banjul is located at 13°28' North, 16°36' West (13.4667, -16.60). In 1816, the British founded **Banjul** as a trading post and base for suppressing the slave trade. It was first named *Bathurst* after Henry Bathurst, the secretary of the British Colonial Office, but was changed to Banjul in 1973.

En route, we were accosted by a young man who wanted to be our guide and followed us down to the market. Here, he intruded on to the turf of another man who *also* wanted to be our guide. For a while, it looked like a fight would break out, but things seemed to simmer down. The second man showed us the day's catches, dumped in rusty non-working freezers with piles of dirty ice cubes. Fish included barracuda, sole, and butterfish.

At this point, Lisa was approached by a man from the hotel who warned her that that the men we were talking to were dangerous, and that we should go. Lisa warned me of this, but I was too intent on asking them about `Gambo`, and thought, because there were three of us, that we would be OK.

I showed them the picture of `Gambo`, and asked them if they had ever seen anything like it. They identified the drawing as a sawfish – peculiar animals related to sharks and rays. Their most striking appearance is a long, toothy snout. They possess a cartilaginous skeleton and no swim bladder. They are the sole family Pristidae of the order Pristiformes, from the Greek and Latin *pristis* meaning "sawfish" (cf. Greek πριστήρ *pristēr* meaning "saw").

They took us on to show us were the fish were cured and dried. We were also shown a large dish of catfish swim-bladders that were exported to China, apparently to help in the production of plastic!

After the tour, the second man asked if we could buy him a bag of cement to help repair the room where the women cured the fish, and as he was not asking for money for himself, but for something to help the community, I agreed.

I asked both men about Ninki Nanka. The first said that one had killed a man from his village. Apparently, the man had been out with his dog, when the animal had started to bark furiously at some bushes. On investigation he found a Ninki Nanka coiled in the undergrowth. It had a face like a kangaroo and a forked tongue. The man ran back to the village, but died half an hour later. The dog, however, was unhurt.

The second man said that Ninki Nanka was like a snake, covered in scales that were like a mirror in which you could see your reflection. He said it could grow as big as a palm tree, and as an example he pointed out some large palms about 60 feet tall. He went on to say that, at first Ninki Nanka was small, but when it grew big it moved into the sea. I have heard similar stories of snakes becoming dragons in the folklore of China and Scandinavia.

The men then led us to a shabby market in order to buy some cement. Lisa refused to enter the unwholesome looking place, and headed back to the hotel. Chris Clark and I followed the men into the market. The first stall we visited was shut, so we headed deeper into the bowels of the market, down malodorous alleys to a stall run by a toothless hag purveying bags of cement.

Just as we were buying the cement, Lisa reappeared with a policeman and another man in tow. One of the men worked in our hotel, and had seen us enter the market with the men, who were known scam-artists and he had phoned the police. A policeman had then caught up with Lisa and another man had shown them the way we had gone.

The policeman told us that the second beach-bum had been known for muggings, and was dangerous. Apparently, he would be getting kick-backs from the sale of cement that probably never went anywhere close to the fish market.

As we returned to the hotel, the police officer left. The minute he did so, the man who had shown him where we were, started a scam of his own. He invited us back to his house, claiming he had just got married. When we refused to go, he asked for a wedding gift of money, thrusting a paper and pen into our hands for us to write down how much we were going to give him. Lisa did an Oscar-winning rendition of having an upset stomach, and we returned to our hotel.

 I felt angry, both with myself for having been so stupid, and also with the bums for exploiting my good nature in such a cowardly and underhanded way.

That night, Lisa *did* actually come down with a case of Banjul belly.

Day 5, 8th July

We rose at 5.00 am for the long trip to Kiang West National Park, the mangroves therein being a reputed lair of Ninki Nanka. Lisa and Chris Moiser were staying behind due to illness. Hassan arrived with a driver called Kamara, who had a four-wheel drive, open-backed Landrover. We picked up Bakary at Abuko as he knew the way, and could also show us the crash sight where the furrow caused by a Ninki Nanka had forced a lorry off the road.

About 4 km outside of Baku, the tarmac roads ran out and the trail became dust. Occasionally, the tarmac would return, but it was so pot-holed that Kamara would drive in the dirt rather than on the tarmac. The landscape here alternated from cultivation to forest and scrub.

Sometimes, we would stop at a village and hordes of children would rush out and scramble over the Landrover with cries *of "Bhaaa, give us sweets", "Bhaaa, buy us a ball"* or even *"Bhaaa, give me thirty dollars"*. On one occasion, a girl of about seven said to Oll, *"If you give me sweets I will give you sex when I'm older"*. The whole experience was appalling. [1]

We drove for the best part of four hours before we came upon the crash site. The rusted remains of a lorry lay beside the road about 250 feet from a bridge. The bridge

---

1. EDITOR'S NOTE: Oll has asked us to stress that he refused this kind offer in a gentlemanly manner.

had supposedly been erected over the great furrow the dragon had left, as it crawled through the rain-soaked dirt that fateful night. The remains of the lorry had been cannibalized, and the rust on it was very advanced. I don't know much about the effects of the climate on metal in the tropics, but the wreck looked as if it had been lying there from before 2001; the earliest stated date of the crash. Oll pointed out the advanced state of the saplings growing through the wreck, where some had hardened into true wood.

We looked at the bridge, and could see that it was, in fact, a drainage culvert. The erosion on the concrete made it look older than five years. Also, I could not envision a creature, however big, making a furrow big enough for me to walk through, just by crawling along. The story didn't add up, and I fancy that the Ninki Nanka is a big scaly scapegoat; a bogey-man, on which accidents and other bad happenings can be conveniently blamed.

On the edge of Kiang West, we stopped at a small village and picked up a man named Bula, who would lead us to the dragon's lair. Assan and Kamara stayed with the vehicle, whilst the rest of us tracked out through the sparse forest towards the lake. The lake, known locally as Yesyes, was more of a flood-plain for the River Gambia. Due to the lateness of the rain, its levels were very low and most of it was a blinding white of salt flats. We saw the tracks of baboon, hyena, and small antelope. Suddenly Bula became frightened, stopped dead in his tracks and refused to go any further, hiding behind some bushes. Bakary translated as Bula told us that he was too scared to approach the mangroves at the far end of the lake. He said that his people had used parts of the lake for cultivation for years, but they dare not go near the swamp because of the dragon.

There had once been a village closer to the swamp in his childhood. After his uncle had seen the dragon, the villagers panicked, the place had been abandoned and it had long since fallen down. A foreign company had tried to build a picnic area for tourists on a point beyond the swamp, but the money had run out and the buildings were left half-finished. We walked further along the lake-bed, until we reached the edge of the swamp. Here, Bakary's nerve left him, and he stayed behind, as the four of us pushed on into the thick mangroves, where fiddler crabs scuttled from our path, and the mangroves grew around huge blocks of dark volcanic rock. At this point, we came upon the tracks of Cape clawless otter (*Aonyx capensis*). [1]

The swamp had an eerie silence, and it was easy to see how it could have got an evil reputation. Malaria bearing mosquitoes in the mangroves may have caused many deaths, and the dragon legend was a way of keeping people away from the area.
Finally, the swamplands began to thin and we saw the river. It was about a mile wide at this point and mudskippers lolloped down its banks. It was obvious that not *all* people were afraid to come here, as there were the remains of a fire and a pile of oyster shells. We saw the abandoned huts of the picnic area; door-less and window-less, looming, skull-like in front of us. I inspected them in the hope of finding bats or

---

1. Although four species of otter are found in Africa, this is the only species known from The Gambia

snakes, but had no luck. We sat and ate our packed lunches in this empty place, the supposed haunt of a dragon. If the Ninki Nanka was here he didn't show himself.

We trekked back through the mid-day sun like the Englishmen (plus one Englishwoman and one Welshman) we were, to where Bakary and Bula were waiting for us. We returned to the Landrover, and drank heartily from the icebox that Kamera had brought along. We dropped off Bula, and then went to pick up another man named Momomodu, who was a park ranger. On the way, a troop of common baboons (*Papio cynocephalus*) emerged from the trees to drink in puddles that had formed in the rough road. Momomodu claimed to have seen the Ninki Nanka less than three years ago in Kiang West. He told us that he saw the dragon emerge from a hole in the ground, and watched it for more than an hour as it crawled around. He showed us some holes that he said the dragon had made. They looked more like the holes of a rabbit-sized animal.

Via Bakary, I put it to the witness that the holes were too small to have been made by Ninki Nanka, but Momomodu replied that earth had fallen into them since the time of his sighting. He showed us another partially-collapsed hole about the size of a manhole.

As I was getting ready to interview him, the Gambian equivalent of the Loch Ness hoodoo [1] struck, and my camera battery ran out. Luckily, Lisa had lent me her Dic-

---

1. **Loch Ness Hoodoo.** This is a syndrome which occurs over and over again in the annals of forteana and it is one that is most frustrating to the investigator. Again and again a potentially valuable piece of film doesn't come out, is over exposed, is double exposed, is stolen, or lost. Ted Halliday described, at some length a similar syndrome which besets monster hunters world wide:

"..........it was clear that the Morar monsters were acting in exactly the same curious way as the ones in Loch Ness. Either a camera was not available to record what was observed or, if it was available, circumstances frustrated the photographer. Almost everyone rejected such a notion because it introduced an element of irrationality. It also raised doubts about the true nature of dragons which those who were anxious to press the claim for an unknown animal chose not to encourage. Normal animals do not behave in such an inexplicable way because they cannot; therefore you had to conclude that the peculiarities were due to chance. This was the prevailing attitude amongst the investigators.

An explanation based on chance seemed to me most unsatisfactory. Chance is a random effect; it is just as likely to work in favour of the investigator as against him. If the ten years of intensive effort at Loch Ness which resulted in failure to get a detailed film was the result of chance then it was not a random effect and the expression became meanmgless. In that event, the explanation lay elsewhere."

Over the next few years Holiday proceeded to collect a large number of "examples of what appears to be some sort of a mental block in relation to the phenomena. There is a desire to minimize or dismiss what one has seen and this provides a brief interim in which the object escapes further observation. This included testimony from such people as naturalist and author Gavin Maxwell, and led him to believe that this syndrome was somehow part of the nature of what he was studying. When he had come to this realisation it was but a short step to another series of revalations:

"Even more marked than the above were the situations in which monster phenomena seemed to be actively evasive. The Loch Ness Investigation Bureau's camp at Achnahannet, for example, has maintained a watch over this part of the loch for six months of the year every season since 1965. But no major sighting has taken place in this area since the Bureau set up its 35 mm. cameras and 36 inch lenses. Yet in 1964 there were two authentic views of monsters from this spot by multiple witnesses which I have interviewed.

The apparent evasion of Bureau cameras by the phenomena is of long standing. An early case occurred in 1965 when a camera located on a platform near the Clansman Hotel was taken away for servicing. The next morning, the staff of the hotel had dramatic views of a hump moving about inshore. Over the years, quite a catalogue of such incidents built up!"

taphone.

This was the only first hand witness we had come across so far, but his description was one of the most fantastical. He said the Ninki Nanka had a horse-like face and was covered with scales that reflected like mirrors. The colours he recalled were mostly green and black, and it bore a crest, "like feathers", that hung down across its face. Its claimed size was staggering - 50 meters long by a meter wide; in other words around 160 feet long. Back in Thailand, in 2000, police officer Suphat claimed to me to have seen a Naga of 60 metres or 190 feet long swimming in the Mekong. This, I had reasoned, was a big animal pulling a long wake behind it, creating the illusion of even greater length. However, Momomodu's sighting was entirely on land, where no such confusion was possible and the idea of such a titanic beast crawling around undetected, was hard to swallow. I asked Momomodu if he saw the creature's teeth, or if it spat fire, and he replied – quite reasonably - that the creature had kept its mouth closed so he had not seen its teeth, and it did not breathe fire. He added that it had neither legs nor wings and resembled a titanic snake.

I showed him pictures of a Nile monitor, a Komodo dragon, an Apatosaurus and a Chinese dragon. I asked him to pick the one closest to what he saw. He pointed out the dragon, saying that the head was very like that of the Ninki Nanka.

His sighting had lasted for an hour. About two weeks afterwards, he fell sick and lesions began to form on his skin.

He went to see an Imam, an Islamic holy man, who knew that he had seen a dragon, and that its baleful influence had caused his illness. The Imam brewed a herbal potion that cured him. Momomodu also told me that if a black man looks upon the dragon he dies, but that it has no effect upon a white man.

What were we to make of this Brother's Grimm-style [1] story? Momomodu said he had found some scales from the dragon, and was willing to let us see them to back up his story. However, time was getting on, and we had a drive of over four hours ahead. We wanted to be back on decent roads before nightfall, so we had to decline his offer and press on home. Something told me that I would not be impressed with his "scales" anyway.

Bakary told us that six or seven years ago there was a white man working for the Parks and Game Department in Gambia. He had been based at Kiang West and had seen a Ninki Nanka, but Bakary could not recall his English name, although the locals called him "Sudokodo". We decided to ask the people at Abuko if they had heard of him.

---

1. The Brothers Grimm (*Brüder Grimm*, in their own words, not *Gebrüder* - for there was a third brother: Ludwig Emil Grimm, the painter) were Jacob and Wilhelm Grimm, German professors who were best known for publishing collections of authentic folk tales and fairy tales, and for their work in linguistics, relating to how the sounds in words shift over time. Jacob Ludwig Carl Grimm and Wilhelm Karl Grimm were born in 1785 and 1786, respectively, in Hanau near Frankfurt. They were educated at the Friedrichs-Gymnasium in Kassel and later both read law at the University of Marburg.

Ninki Nanka - A Giant Monitor Lizard?

Day 6, 9th July

We decided to take it a little easier after yesterday's mammoth journey. 90km is nothing in Britain, but my familiarity with foreign roads had taught me that such a trip would take an age! We made some plans for what to do next, and Lisa came up with the excellent idea of checking Banjul museum to see if they had anything historical on Ninki Nanka. We also made plans to visit a sacred crocodile pool in the south of the country, along with a reptile park, where two specimens of the ultra-rare Armitage's skink had been held, and also the Alaheen River that lay between Gambia and Senegal.

Day 7, 10th July

Banjul was far less hassle where beggars are concerned than Baku (where our hotel was). There were a few hunchbacks and lepers with begging bowls, but no-one followed you around or became intrusively. Some of the lepers rode hand-wheeled bicycles.

As Chris and Lisa were both feeling better, they accompanied us once again. Chris Moiser wanted to track down a song about the Ninki Nanka, so we went to a market place first. A man at a music stall told Chris he had a copy himself, and could make a copy for him. Chris was taken to the man's flat, with several other men, whilst we waited in the market below. As time drew on we became increasingly worried about his safety, particularly after what had happened at the fish-market. Finally, however, he emerged with the tape.

We moved on to the National Museum. This was a truly "Gambia-style" affair. Looking as if no-one could be bothered to lift a finger for its maintenance, it was very 'down at heel', and included some archaeological exhibits on the pre-history of the country, some basic natural history, some tribal exhibits, and quite a bit on colonialism. There was nothing about Ninki Nanka *per se*, but there was some information on a folk belief from Ghana. The story ran that an ancient king had made a pact with a giant snake that brought the rains. If they fed it on a diet of young girls, then it would provide good rainfall for the crops. The snake then took up residence in a well, but the king's son fell in love with one of the intended victims, and fought the giant snake and slew it, thus freeing the girl to marry him.

Those who study folklore will see several re-occurring motifs here:

1. the "dragon" in control of rain, and dwelling in a well
2. the female sacrifice offered up to the dragon,
3. and the noble hero who saves her.

This story may have influenced Gambian beliefs.

Chris Moiser asked the staff if they knew anything about the Ninki Nanka, but they knew almost nothing, and thought that the belief had now largely been forgotten. It seemed like they had not done their homework. Indeed, the staff seemed unfriendly, lazy, and very unknowledgeable about anything except taking money off visitors.

We visited a colony of epauletted fruit bats (*Epomophorus spp*) that roosted behind the impressive Atlantic Hotel. After watching the snoozing bats, we toasted Jon Downes' late father, J.T. Downes, before returning to our own hotel.

On the way, I asked our driver Assan if he knew anyone who had seen the Ninki Nanka. He said that he knew an old man in Guinea who had seen it when he was a boy, and who was now a grandfather.

Day 8, 11th July

Oll's birthday.

We had intended to visit the south of the country today, but Assan could not make it until tomorrow. We had a crushingly boring day at the hotel catching up on our notes. I am enjoying this trip far, far less than Sumatra, or Mongolia.

Day 9, 12th July

After what I considered to be a wasted day yesterday, we filled this one with activity. We headed to the south of the country *sans* Chris Moiser, who was suffering from heat exhaustion. Assan, and another driver called Omah, drove us the 40km or so down to the Alahein River. We paid some fishermen to take us a mile or so down river in dug-out canoes. We were going to visit an island where pelicans and fruit bats were resident, but at high tide the island was little more than tree tops.

Instead, we crossed over to the far bank into Senegal. We walked for about half a mile into Senegal, observing mudskippers and fiddler crabs. There were piles of shells left beside the river and Assan explained that they were used to create white-wash. One large pile had a black, forked stick with black beads attached and Assan said it was juju to prevent anyone stealing the shells.

I asked the boatmen, one of whom was from Mali, what they knew of Ninki Nanka. The Mali man, who, like the man from Kiang West, was called Momomodu, said that it could transform itself from a tiny snake into a gigantic one. The other boatman, who was from Senegal, echoed his story.

Moving on, we visited the Sacred Crocodile Pool at Katung. The water was low, but the guide told us that there were around twenty crocodiles in the pool. They had not been brought there, but had always existed in these naturally occurring pools. We saw only one young crocodile; a yearling by the look of it. It was good to know that they were breeding. The guide told us that women who wanted to conceive visited the pool for its fertility magick. There was also supposedly a white crocodile in the pool, and to see this reptile was considered great luck.

When we asked him about Ninki Nanka, the guide said that it was a snake that grew into a huge python then, in turn, went into the sea and became a dragon.

There was a large group of very well-behaved children at the pool. The eldest was a youth of about fifteen, and told us that a song about Ninki Nanka was traditionally sung after male circumcision. After circumcision men do not bathe for one month, after which time they go down to the river to wash and sing the Ninki Nanka song. He then proceeded to sing the song for us.

Oll came up with the idea that the Ninki Nanka is a vestige of a pre-Islamic Animist religion and could, perhaps, be some python deity that has become demonised.

Finally, we visited the Gambian Reptile Park where there is a collection of snakes, lizards, chelonians and other Gambian wildlife. The director was a likeable Frenchman called Luc Paziand, who had been living in the Gambia for twelve years and running the park for four. Luc removed problem snakes from people's properties and kept them at the park to educate the public. He was also working on a new anti-venom for the puff adder.

Luc was delighted to meet some fellow scientists and invited us in for a drink. His youngest son was clambering everywhere like a monkey. At 18 months he had no fear and luc told us that he had found him up on the roof only a few days earlier and that once he had also found him teetering on the edge of a sixty foot deep well!

He had kept specimens of Armitage's skink [1] until only the week before, when a pair had been captured by a local man in the toilet of his bar at Gunjur Beach. One had been injured and had died, but luc had kept the other for some time before releasing it, when he found it difficult to find a regular supply of mealworms for its diet. He told us that the best place to look for Armitage's skink was the grassy dunes behind Gunjur Beach.

He told us that the Mandinka name for the skink was 'bauko kono sa'. It translated as

---

1. Armitage's Skink (*Chalcides armitagei*). Chris Moiser wrote before the expedition:

"Named after its discoverer Captain C. H. Armitage by E. G. Boulenger then head of reptiles at London Zoo who first described it in 1922. To look at it is a fairly unremarkable small lzard about 5" long with a tail of about 3.5". Unlike most of the other lizards in the area where it comes from it has 3 toes on each limb. Captain Cecil Hamilton Armitage (1869-1933) was the Governor of the Gambia from 1920 – 1927 and was described as a generous donor to the Zoological Gardens of London.

The 3 (living) specimens of this lizard having arrived in a collection of reptiles that Armiatge had sent from the Gambia. These specimens ultimately ended up in the British Museum (preserved), where 2 are labelled "Gambia" and one "Cape St Mary" In 1989 the lizards were effectively rediscovered by The Gambian Dwarf Crocodile Project people. They believed it to be restricted to a narrow sandy coastal fringe which was rapidly being lost due to the increase in tourist hotel construction. The site where the found the animals in 1989 had been cleared for construction in 1990. An FAO survey suggests that they may also exist in the Tanji area".

'snake inside the sand'; a reference to its snake-like body with tiny legs and its ability to disappear swiftly into the sand.

We asked him about Ninki Nanka and he repeated the, by now, familiar story that the locals believed it was a snake that grew into a huge python, and then entered the sea to become a dragon. Luc thought that perhaps people had seen large pythons with ticks or leeches attached to their heads and folk mistook these for a crest.

Luc was thinking of packing up and leaving Gambia, for the less-disturbed and less-explored Guinea.

On the journey back through Katung, Assan took us to meet the grandson of his friend who had seen a Ninki Nanka in Guinea. The man was called Lamu and his story was quite fantastic. He claimed that his grandfather could call Ninki Nanka and pacify it with gifts of fish and rice. He also claimed that his grandfather could travel to Gambia and perform the same feat for us, if we were ever to return.

Lamu said he had never seen his grandfather tame Ninki Nanka, but said it looked like a huge snake, but with a round head. This jarred with the descriptions of the beast having a horse-like head. He showed us an alleged scale from the beast, which looked more like a chip of old glass. Altogether, I was not impressed by his story.

Day 10, 13th July

I did an interview with BBC World Service, (Africa), about our expedition, during which the usual questions were asked. What are you looking for? What is the Ninki Nanka? What evidence have you found? They even asked if the trip was just an excuse for a holiday. I explained that our team consisted of a zoologist, a biologist, an ecologist, an archaeologist, and even an astro-physicist.

Later, we travelled down to Gunjur Beach to look for Armitage's skink where we stopped by a small beach bar for a drink of pop. It was a home-made building, consisting mainly of driftwood and shell wind-chimes that jangled in an unearthly manner in the breeze. It turned out to be the very bar luchad told us about. The owner described the lizards he had caught as being about ten inches long, as thick as his finger and sandy brown with a dark stripe. He had captured them in his toilet and handed them over to Luke. Odd indeed, that two such rare specimens turned up in an outside toilet!

We started out along the dunes. It was our plan to search, just as it was cooling down in the late afternoon, when reptiles and their prey would be more active as the fierce heat of the day subsided.

As we walked along the dunes, a man and his dog approached us totally uninvited. He didn't say a word, but just followed us around like a child. His dog was snuffling around in the undergrowth destroying any chances of us finding a skink. Suzi pointed out that we didn't need any help, and that his dog was frightening away the lizards we were looking for. He ignored her, so I asked him quite sternly why he was following us. He bleated that he was trying to help us, but I knew enough about Gambian beach-bums to know that "helping us" equalled to getting hold of our money. I firmly explained that he was getting in the way, and asked him to leave. He shuffled off with his dog in tow.

We scouted along the dunes, then turned back as the light began to dim. To our horror, the irritating man came into view again, this time without his dog. He began to follow us once more, and claimed to have seen a snake beside a road here four years ago. His description of the snake in no way resembled Armitage's skink. We decided to give the search up as a bad job and call it a night.

On the upside, we had confirmation from two independent sources that Armitage's skink is still extant and we have increased its known range by around thirty miles.

Day 11, 14th July

Lisa's birthday.

Today we visited the Palma Rima Hotel. Chris Moiser had heard a story that a Ninki Nanka had lived in a hole in the area, prior to the hotel's construction in 1990. The story goes that the locals blocked up the hole while the beast was away and planted a tree over it to stop the Ninki Nanka returning. The hotel was then constructed. Assan recalled the tradition of a Ninki Nanka living in a tree in the area. Perhaps the story of a dragon in a hole was a confused re-telling of the story.

Before we even got to the hotel, the first of the beach bums descended in search of money. He said, *"Remember me, I am from the security at your hotel."*

"And what hotel is that?" I replied

"Palm Beach Hotel".

"I'm staying at the African Village".

"Oh, I'm your driver's nephew".

"What is his name?"

"Bula".

"No, his name is Assan."

At this point he knew he was beaten, and sloped off.

The hotel was closed, and there was no-one we could talk with to see if they could recall anything. Chris thought a large baobab tree was the one with the story attached to it, but it looked far too big and old to have been transported to the spot in 1990.

The tree did have *some* significance, however, with what looked like prayers or spells having been hammered into the bough with wooden pegs and bolts. Maybe some old story of a Ninki Nanka associated with the tree had been corrupted into a modern day fable.

We returned to Baka Samba's shop, where I had hoped to interview him further, but he was away. Lisa and Suzi bought some silver jewellery before we moved on.

We had been studying the maps more closely, and thought that the carcass of `Gambo`, if it was still in existence, was just about beyond the nightclub walls. The target area was a small patch of sand between a palm tree and the wall of *Destiny's* Nightclub.

By this time, we had attracted a small crowd of beach-bums who seemed to have crawled out of the woodwork, along with a local juice salesman, and a couple of people from the nightclub (which doubled up as a restaurant by day).

Deciding that our presence was less likely to be questioned if we had the appearance of officialdom, we told them that we were a group of geologists who had been sent by a British university to investigate beach erosion in the Gambia, and also that we were worried that Bungalow Beach would erode in the same way as the beaches at Baku. [1] They all fell for this explanation, and seemed genuinely interested - our digital camera and Chris Clark's GPS device looked sufficiently high tech to fool them. They didn't seem to notice that this 'university-backed' geological expedition was only equipped with children's plastic spades, more usually used for building sand castles.

We excavated two deep holes in the sand, and it swiftly became apparent that, even at quite a modest depth, the sand became wet, which would have destroyed any remains after 23 years. Also we found that there was much disturbance. Chunks of concrete from the construction of the nightclub were discovered, as well as more recent items such as drinking straws. Whether it was a dolphin or not, `Gambo` would now be long gone.

As we dug, Chris Moiser kept the locals enthralled with an impromptu lecture on beach erosion and its causes. After we had filled the holes in, we assured the onlookers that the beach was fine and that there was no erosion. They were happy and relieved at this news. We enjoyed a drink of mango juice, and reflected that it had been nice to bullshit the bullshitters for once.

---

1. EDITOR'S NOTE: here, I would like to stress, that kying through ones teeth in order to bamboozle gullible natives is not, and will never be, official CFZ policy. However, sometimes one has to adapt to the needs of a particular situation, and especially considering the increasingly delicate political situation in the country, the expedition team had no logical alternative. Anyway, when in Rome….

Back at the African Village Hotel, the reception manager, Bubulha, who had heard my radio interview, had some information for us. He said that, as a boy in the late 1960s or early 70s, he had heard a radio broadcast warning people to stay away from the Fajara area. A Bishop Moloney had seen a dragon crawling out of a rocky area of standing water (close to the postal station) and entering the sea. It left a furrow in its wake. Bubulha also said that he knew an old man, in his youth, who had seen the Ninki Nanka. When the old man was a boy he and a friend saw the dragon. His friend fell ill, his hair fell out then he died, but Bubulha's informant was unaffected for some unknown reason.

Day 12, 15th July

At breakfast, a woman from the reception told me I had a 'phone call. A woman named Mariama was on the line, and she told me that she knew a man who knew were to find the dragon. Apparently, it lurked in a hole in some mangrove swamps beyond the village of Mandinari, and the man said that if you threw a dog down the hole the dragon would emerge to devour it! I stressed that as we all adored dogs, the sacrificing of one to a dragon was out of the question. We would, however, like to visit the hole and maybe throw down a chicken. Failing that, I thought, we could try to smoke the beast out. Mariama said the man did not want paying! I told her that we would meet him at 11 o'clock at Abuko Park the following day. I thought that maybe a huge python was living in the hole, but we needed to check it out and it would make a thrilling climax to the expedition.

Later in the day we interviewed a security guard named Sueliman. He told us that his mother had seen a Ninki Nanka at a place called Upper Niumi when she was a girl. She had been out looking for palm nuts when she and a friend had come across the dragon coiled up in a hole. It had a crest like a fowl on its head that hung down, and this crest seemed to have writing upon it.

At the same time a hunter who was up a tree saw the monster. Sueliman's mother did not speak of the sighting, but the hunter talked freely of it and subsequently died.

Sueliman said that the dragon's crest was supposed to have words from the Qur'an written upon it and that if you read the words you died. Also, if you spoke about the sighting to too many people you would also die.

Perhaps the 'words' on the crest refer to some kind of veining or patternation. Arabic writing lends itself to naturally occurring patterns due to its flowing, rounded, nature (unlike the straight edged western lettering). Ergo, the name of Allah is often found in the pattern of aubergine seeds and in other fruit and vegetables.

Sueliman offered to guide us to the dragon's lair, but would not approach it himself. However, the area was far away on the north bank and we simply did not have time to go there.

At dinner that night, I was phoned directly by the man who claimed he knew where the dragon of Mandinari lurked. He seemed almost frantic and very aggressive. He said he could lead me to the dragon and bait it out with a dog. I said I would not use a dog, but would use a chicken instead. He then began to shout *"What do I get? What do I get?"* (like the old punk song [1]) in an aggressive manner.

He shouted at the people in reception to be quiet. Then he said he would be risking his life to show us Ninki Nanka and that he wanted to be paid. *"£3000, £4000, £5000 how much do I get?"* he bellowed. This was in pound sterling, *not* Dalasi! I declined his offer. We didn't have that kind of money and besides he was an obvious chancer who had heard the radio interview and was suddenly an expert who knew were to find the monster, for money of course!

We were going to Abuko tomorrow anyhow, so we decided to have a look at Mandinari on our own.

Day 13, 16th July

Assan took us to Mandinari, where, beyond the tatty little village, lay a huge expanse of mangroves [2]. A rickety jetty of flimsy looking logs stretched out into the swamp. Suzi and Chris Moiser walked out onto it for quite a way, but it began to give under

---

1. *"What do I get?"* was a single by Manchester punk band *The Buzzcocks*, usually accepted as the first punk band formed outside London in the wake of the *Sex Pistols*. The single was released in February 1978, and reached #37 in the charts.

2. **Mangroves** are woody trees or shrubs that grow in coastal habitats or *mangal* (Hogarth, 1999), for which the term **mangrove swamp** also applies. Mangrove plants are found in depositional coastal environments where fine sediments, often with high organic content, collect in areas protected from high energy wave action. Mangrove plants are a diverse group which have been able to exploit a habitat (the intertidal zone) because they have developed a set of physiological adaptations to overcome the problems of anoxia, salinity and frequent tidal inundation. Each species has its own capabilities and solutions to these problems; this may be the primary reason why, on some shorelines, mangrove tree species show distinct zonation due to variations in the range of environmental conditions across the intertidal zone. Therefore, the mix of species at any location within the intertidal zone is partly determined by the tolerances of individual species to physical conditions, like tidal inundation and salinity, but also may be influenced by other factors such as predation of their seedlings by crabs.

Once established, the roots of the mangrove plants provide a habitat for oysters and help to impede water flow; thereby enhancing the deposition of sediment in areas where it is already occurring. It is usually the case that the fine, anoxic sediments under mangroves act as sinks for a variety of heavy (trace) metals which are scavenged from the overlying seawater by colloidal particles in the sediments. In areas of the world where mangroves have been removed for development purposes, the disturbance of these underlying sediments often creates problems of trace metal contamination of seawater and biota.

It is often stated that mangroves provide significant value in the coastal zone as a buffer against erosion, storm surge and tsunamis. While there is some attenuation of wave heights and energy as seawater passes through mangrove stands, it must be recognised that these trees typically inhabit areas of coastline where low wave energies are the norm. Therefore their capacity to ameliorate high energy events like storm surge and tsunamis is limited. Their long term impact on rates of erosion is also likely to be limited. Many river channels that wind through mangrove areas are actively eroding stands of mangroves on the outer sides of all the river bends, just as new stands of mangroves are appearing on the inner sides of these same bends where sediment is accreting. They also provide habitats for wildlife, including several commercially important species of fish and crustacea and in at least some cases export of carbon fixed in mangroves is important in coastal foodwebs. In Vietnam, Thailand, the Philippines, and India, mangrove plantations are grown in coastal regions for the benefits they provide to coastal fisheries and other uses. Despite replanting programs, over half the world's mangroves have been lost.

my weight so I decided to stay on the edge with Lisa. I spoke to a local youth called Mustapha and he told me what his mother had told him about Ninki Nanka. He said it was a huge animal covered in scales like mirrors. On its head was writing, and if you were wise enough to read the words you would live. The Ninki Nanka could also read the words written on human heads (thoughts?).

After telling me this, Mustapha began to whine for money like a child whining for sweets. Apparently, he also grasped Suzi's hand and told her that he loved her. We thought it was high time to move on.

We returned to Abuko and enquired about the Englishman named 'Sudokodo' by the locals. [1] No-one could recall such a man so we set off for another jaunt around the park where, at the pools, we saw eleven baby crocodiles. It is clear, therefore, that these reptiles are breeding in at least two parts of the Gambia.

Without invitation a rather poor guide, who was not anything like as good as Musa, unfortunately attached himself to our team. He did not know anything about the animals and made an awful lot of noise. The guides are paid out of the admission price, but unlike Musa, the uninvited guide wanted 250 Dalasi! (about a fiver). He also wanted Oll's watch. We did not oblige.

---

1. EDITOR'S NOTE: When they first tried to contact this man, they sent me the following email request, which I posted on the blog:

"Please put out the following information through any avenues you can.

We're searching for an Englishman who was attached to, or working with, the Gambian Wildlife Authorities in the Kiang West National Park area in the 1996-2002 period. He may have had the local name of Sodokudo, or something similar. Clearly, if he was a researcher, he may in the meantime have returned to the UK, and/or moved onwards. It is believed that this man may have had contact with a Ninki Nanka and as an eyewitness would be of great assistance to us. Any information as to his identity or how to contact him would be gratefully received."

The next day I received the following email:

Hi John

Having asked around in The Gambia our resort manager has sent the following:

There was an Irish Man - Paul Murphy who worked with the Wildlife at Kiang West he was with VSO - who are no longer here but have an office in Dublin so you may be able to trace him through the Irish Office. Here say - They say he was a bit of a lad, good fun and could well have told the Ninki Nanka to keep the locals out of the park!!

There are always Peace Corps out there too but they are normally from the States,

Sorry, not sure if this is of any help. I do hope so.

Kind regards

Jo Wedeman
Marketing/PR Executive
Serenity Holidays
www.serenity.co.uk

At the time noted UFOlogist, and sleuth Nick Redfern was staying with us, and so I set him on the task of hunting for Mr Murphy. However, by 'Murphy's Law' (if you excuse the pun), no-one at the VSO offices in either London or Dublin had heard of him. Neither had the Peace Corps, so we sadly had to give up.

After leaving Abuko, we stopped at a little bar run by a friendly Rasta named Max. Chris Moiser had met him before, during the days he brought students to the Gambia. Max offered to lead us to areas of Ninki Nanka sightings for free if we ever returned. He was appalled at the idea of people asking for money.

Max had not seen the dragon himself, but once, as a child, he had an odd experience at a creek in Bintang Bolon. He almost drowned and felt 'electricity' in the water and refuses to as much as wash in the creek to this day. He also told us that, in the 1970s, Eddy Brewer, the Englishman who had turned Abuko into a National Park had once erected mirrors around the lakes there.

Day 14, 17th July

Back at the hotel, I received a note from Joe Miuleh, taxi co-ordinator, who claimed he could show me Ninki Nanka, but he wanted paying.

We visited the offices of the *Daily Observer*, where we hoped that in return for a story on us, they would let us look at their archives. The offices were very down at heel, and looked as if they were ready to fall apart at any moment. They did not even have a single copy of their latest edition, let alone any kind of archive!

The interview, such as it was, was mainly carried out between Chris Moiser and the agricultural correspondent. We also slipped into an internet café next door so he could download our blogs off Suzi's memory stick. We ended up paying for the internet café.

Back at the hotel, the painting I commissioned from a local artist several days previously turned up. It was supposed to be Ninki Nanka. It was, in reality, a poorly realised generic Chinese dragon, with a background of mangroves and could have been copied from anywhere.

Joe, the taxi co-ordinator, spoke with me and told me that he knew were to find Ninki Nanka. He said that it would only come out at three in the morning on Thursdays, but he could take me to it if I paid him. I declined, and he said he could get me photographs as proof. I left my postal address and fully expect some crude fakes to reach me some day.

Day 15, 18th July

Today we leave. Gambia has not inspired me in the way Mongolia, Thailand, and Sumatra have. Once they had heard the radio interview, every man and his wife seemed to know where to find Ninki Nanka, and the beggars and beach-bums make trying to do research here a thankless task.

As I waited for the coach to take us back to the airport, a man named Momomodu

Ninki Nanka woodcarving

said to me that Ninki Nanka lived on an island up-river. It looked like a mighty snake with multi-coloured scales. He offered to take me to the area in a four-wheel drive if I ever returned.

So what are my thoughts on this beast?

Most people have not seen it for themselves - their mother/father/uncle saw it long ago, and the only first-hand witness gave a description that was very hard to believe. Several said that Guinea was the place to search for Ninki Nanka. I think some form of huge snake may have existed in former times, but is now extinct in the Gambia. It may still lurk in the less well-travelled areas of west and central Africa. The animal has become a bogeyman, and has left behind a very real fear. Perhaps the influence of Islam has demonized this creature from the past.

It is interesting that the descriptions of a giant crested snake are akin to those I was given in Thailand back in 2000 (the year of the dragon). There, however, I spoke with many more first-hand witnesses. I postulated that this could be a species of surviving Madtsoiid, a group of gigantic primitive snakes believed extinct for 10,000 years.

The medieval bestiaries described the dragon as a vast serpent with a crest. They placed them in Africa. Perhaps they were correct. In the less trodden parts of the Dark Continent here, indeed, be dragons.

In the shade of a termite mound (Abuko)

Abuko: Skull of a warthog (*Phacochoerus africanus*)

A primal morass - Abuko National Park

A termite mound in the jungle

Rainbow Agama *(Agama agama)*

Hatchling Nile crocodiles *(Crocodylus niloticus)*

Darwin Centre (Abuko)

aka Samba and Richard Freeman

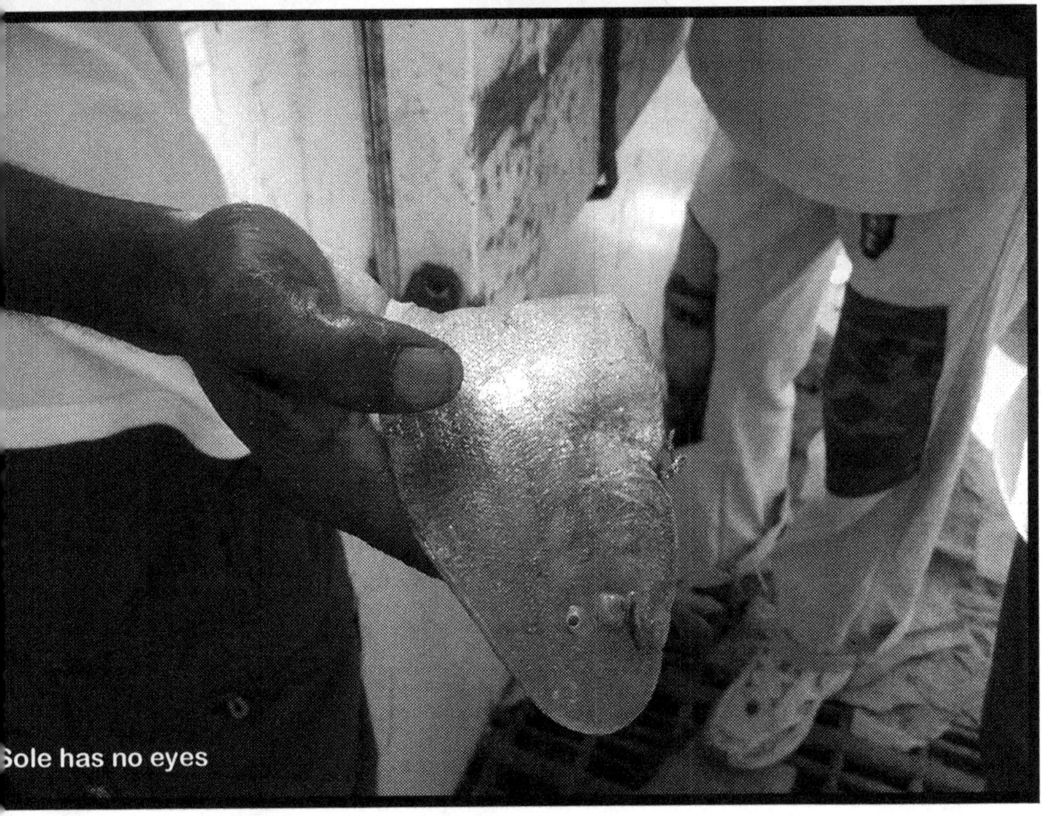

Sole has no eyes

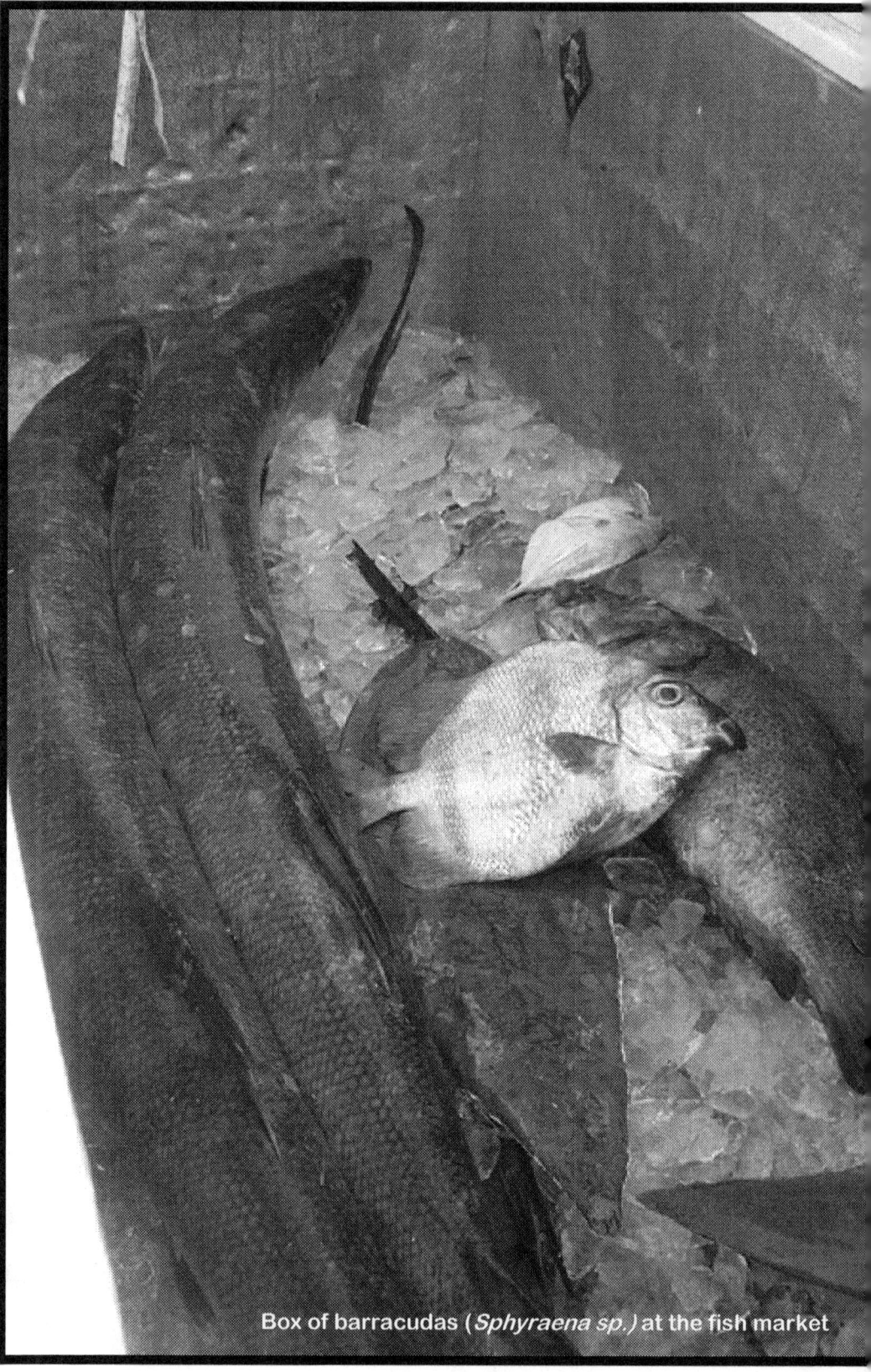

Box of barracudas (*Sphyraena sp.*) at the fish market

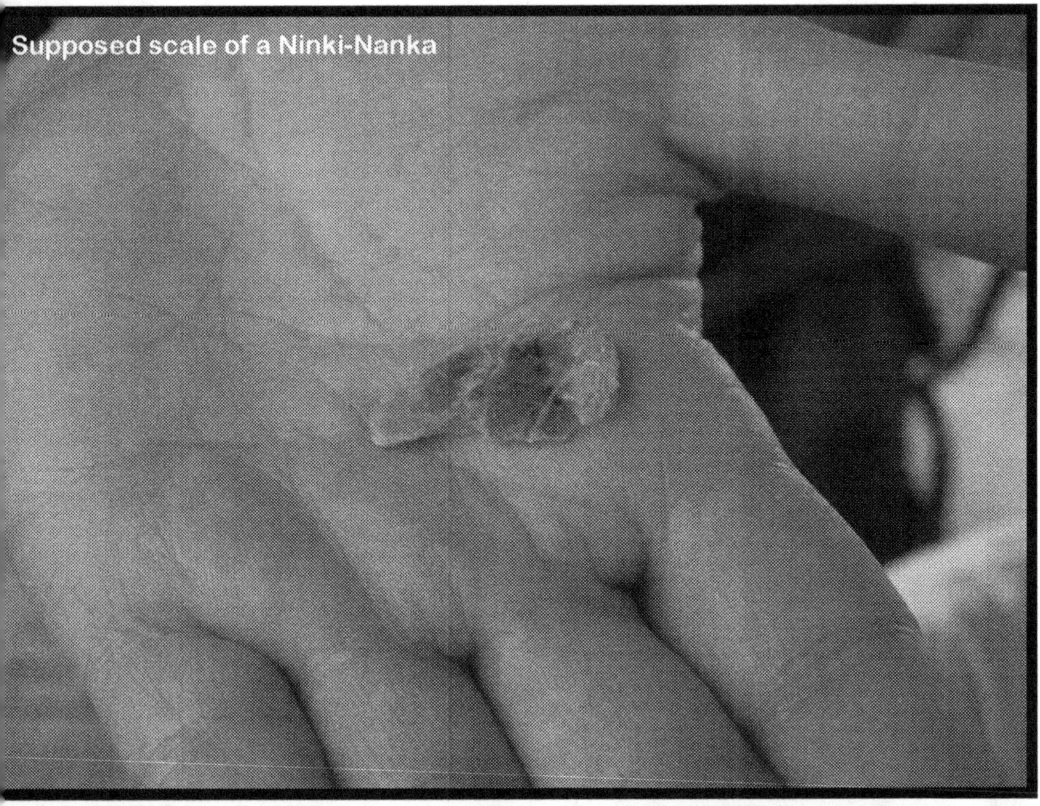

Supposed scale of a Ninki-Nanka

A collection of alleged Ninki-Nanka scales

The culvert over the supposed dragon furrow, Kiang West

The wreckage caused by a dragon encounter?

Dry lakebed at Kiang West

Richard in the dragon's lair

The home of Ninki-Nanka

Baboon tracks (Kiang West)

The abandoned huts at Kiang West

Empty huts beside the swamp

Baboons *(Papio cynocephalus)* at Kiang West

Momomodu, did he see the dragon?

A supposed Ninki-Nanka hole

The Alahein shoreline

Kartong, beside the Alahein

Boats at Kartong

Exploring Senegal

Looking into Senegal

The ju-ju stick

The site of the 'Gambo' excavation on Bungalow Beach

The excavation of the burial site

Chris M. tells the locals about beach erosion

The swamps of Mandinari

Jetty leading into the swamps

Low tide at Mandinari

The mangroves of Mandinari

Interviewing a local at Mandinari

A puff adder (*Bitis arietans*) at the Gambian Reptile Park

Side necked turtle (*Pelusios subniger*) at the Gambian Reptile Park

Painting of Ninki-Nanka by local artist

Chris Moiser

Chris Moiser is a zoologist who taught in a College of Further Education for 20 years, but whose heart is really in Africa.

When teaching he started field trips to The Gambia, as a way of getting students into sub-Saharan Africa fairly cheaply. He used the country as an example of the advantages and disadvantages of tourism, and to teach students a little about mangrove biology and monkey biology. He also used these trips to produce two small scientific papers on crocodiles and the roles of sacred pools in conservation.

He has also visited South Africa, Zimbabwe, Botswana, Zambia, Mauritius and Kenya, purportedly on holiday, but he usually manages to get to the local reserves and to get at least one article out of each trip.

In his spare time Chris writes, and his first novel, *While the cat's away* has recently appeared in print. Whilst ostensibly about the Beast of Exmoor it actually starts in Zimbabwe, and flashes back to affairs there in the days when it was Rhodesia. Chris's earlier books are on big cats in the UK and Sea Monsters of the South West. He is currently working on a book on the Cryptozoology of West Africa.

Monday 3rd of July

> *"The Gambia is one of the most oddly shaped countries in the world: it looks like an earthworm, and fits around the Gambia River like a long, tight, wrinkled sleeve."*
>
> John Gunther, *Inside Africa*, 1955

Having booked at the wrong time to get from Plymouth to Gatwick, it was actually cheaper to fly directly from Plymouth, rather than to get the train. Sometimes these things happen. Plymouth airport is, in a way, like an African airport, it has a small tin roofed terminal with a lot of seats, and is either very busy or very quiet.

The flight is going to be 20 minutes late (general congestion) so check-in is 20 minutes late in opening. Having packed my bag myself, and having left my guns at home, check-in is quick and efficient. The aircraft lands as we are ushered through the search point and kept locked up until we can be boarded. They walk you out to the aircraft here. A Dash 8-300 is the aircraft on this route, a twin-engined short take-off and landing commuter. We go to Newquay, about 11 minutes flying from Plymouth, where we change a few passengers and then move on to Gatwick. A longer flight, and at higher altitude.

Landing at Gatwick is simple enough; we are, though, one of the smaller aircraft there, and are shepherded to the outer parts of the North Terminal before disembarking. Baggage reclaim utilises the same procedure as it does for the bigger aircraft though, except that I am in "domestic baggage reclaim". With my bag, and a newly acquired/borrowed trolley, I head for Gatwick South by the transit system. This is not because the Banjul flight is from there, but because it is a larger terminal and the facilities are better. It also has the railway station there, from which I am to meet the rest of the gang in the morning.

Gatwick South has a few shops, "Gatwick Village" - these are undergoing a periodic rebuild though, so just over half are open. I have a burger and a pint of the most expensive lager that I have ever had in the UK. As the terminal is over-heated it is still value for money. After "dining" I feel more like mischief, so I telephone Jon Downes from a pay-phone and tell him that, due to a mistake at Plymouth, I am now at Jersey Airport with no prospect of catching tomorrow's flight. He is not amused, indeed, he is rather panicky [1], so I quickly put him out of his misery and discuss a few things about the trip.

---

1. A base libel! J.D

Tuesday 4th July

Time to try to get some sleep then. You can't sleep at Gatwick, though, you can only doze fitfully between disturbances. W. H. Smiths stays open all-night, so I buy *The Times* and have completed the Sudoku by 04.00am. I even spend a £1.00 coin at the internet café sending the answer off (and no I didn't win). Another doze, and then search for Dr. Chris, Richard, Suzi, Olly, and Lisa, who should be arriving by train. I wait by the station entrance hoping to catch them, but somehow they evade me and Richard telephones me to say that they are at the pre-arranged meeting point. When I get there, Dr. Chris still hasn't arrived; he is coming by taxi and will be a bit later. We decide to at least get rid of the baggage so we go on the transit to Gatwick North. Olly hasn't been on the transit before and really likes it.

We check in at the Astreus desk, which seems less permanent than the others. Lisa has some large baggage, and that has to be taken elsewhere. Eventually we are "sorted" through, and Dr. Chris has arrived and checked in his bags, so we go through the security checks, and searches to enter the departure lounge. We eat at diverse over-priced establishments and then look at the shops. Our flight seems to be announced (silently though, in the Gatwick way, on the screen) and seems to go straight to final call. In two or three groups we headed to the gate, where in fact not a lot is happening.

At least here we have a sight of the other passengers from the flight. It is the usual mixed Gambia flight. Groups on missions, Church sometimes, research sometimes, families and young couples on holiday, African businessmen returning home, and solitary Europeans. The solitary Europeans are a mixed bunch, there are single males who look tired of travelling (possibly legitimately, as they have probably been up all night), and there are single ladies, these ladies are usually older and larger; one hesitates to use the word sex tourism.... but the idea might just be there, even if they would be horrified at the suggestion.

Boarding proceeds in the usually way, the superior class first, and we suddenly find that Olly is in this group. It appears that he has been upgraded because of the greater seating space available there. The rest of us are then called up. The aircraft is a Boeing 757-200, a nice steady aircraft, well capable of the distance, and a regular on this route. The airline, Astreus, is starting scheduled services to Africa, West Africa in particular. They are not exactly in the BA, SAA or other "national carrier" class, but they are pleasant enough, do their best, and the food is edible. We get a snack and lunch, and cartons of Harrogate Spa water. At least a British airline that gives British water, and not some imported French water.

The flight is uneventful, but a strange co-incidence though, although we were not to know it until after take-off, the first in-flight movie was *King Kong*, the recent Peter Jackson version. Whilst enjoyable, I am glad that I didn't buy the DVD. I, at least, wonder whether there is some divine hint in this, particularly when I see the dinosaur sequences!

In just under six hours we are at Banjul and the fun starts. £5.00 tourist tax for entry (preferably in a "hard currency" not the local Dalasi). Immigration take the forms that we were required to fill in on the aircraft, and stamp the passports with a smile. There is usually quite a bang as they stamp them. In fact, of the nine African countries that I have visited, I think the Gambian's stamp passports the hardest. On to baggage reclaim, where there is the obligatory wait that happens in all airports. Gambia does things differently though, and as the baggage comes through, the porters climb on to the moving belt and grab the better looking pieces.

If they get yours, you either get them to take it to customs where you pay them a £1 coin, or you wrestle it off them. My bag had been to Gambia before so the porters ignored it. (Sadly, travelling to Africa seems to age luggage prematurely). I grabbed it and went to customs. This time they X rayed it on the way in. Mine appears to contain no contraband, at least not recognisable to Gambian customs. Therefore, I can proceed into arrivals, where the *Gambian Experience* representative spots you as his clients, and sends you out of a side door to the coaches. Once you get to the coaches, you have to hand your bag to a man who places it in the belly of the coach and expects £1.00 for doing it. The driver also seems to be magically standing there too, and also needs £1.00 for the service that he may be about to provide.

The *Gambia Experience* representatives seem oblivious to this semi-extortion, and stand some distance back, trying to give you bottles of imported bottled water. This time it was from Tennessee in America! As the afternoon heat at Yundum Airport is searing, the water is almost essential. On the coach, we are given a card inviting us to a welcome meeting. It does acknowledge, though, that there are many regular returnees to Gambia and that you are excused if you do not wish to attend. I do not have a card, so I am allowed to search through the pile; we all appear to be Dr. Clark and party.

The drive to the hotel takes just over half-an-hour. We call at another hotel called Ngala Lodge along the way and drop a few people off there. As we near our hotel, we pass the Medical Research Council Compound, the British High Commission, Bakau Barracks, and Bakau Fire Station - all looking very African in the heat of the afternoon sun.

At the hotel, we are met by an assortment of staff who convey our bags to reception, and subsequently to our rooms, where they wait for the obligatory £1.00 coin wondering whether we will understand what is required of us. (Note to future expedition organisers, give each participant a handful of £1.00 coins before they leave the UK.)

I am surprised in reception to find an old friend, Assan Njie, there. Assan does not work for the hotel, but is a taxi driver and friend from the past. Assan has a past not involving taxis, and he knows that I have an African past not involving tourism. We have a sort of respect, partly based on things we will not tell each other. He says that he did not know that we were coming, but had apparently received a call from "a friend" at the airport. The "friend" even knew which hotel I was going to! It was all a bit suspicious, but these things happen in Africa. I take Assan's card, with his mobile number on, swap some banalities and say that I will call in the next day or so, before heading to the room.

As a group, we have four out of a string of five rooms. These are close to reception and the pool. We will have noise problems, but we have a benefit that most in this hotel do not have, air-conditioning. I close the door for a few minutes of solitude from the hustle and bustle of travel. We have arrived, what will the fortnight bring? As I close the door, the front falls off the air-conditioning unit and narrowly misses my left shoulder. I am back in Africa!

The air-conditioning only takes a few minutes to clip back together, so I have a quick shower and hang up my long trousers, not to be worn now for another two weeks. Shorts on, and a shirt; then off to reception to exchange some travellers' cheques. The hotel give 49 Dalasis to the pound, about right. The notes come in 5, 10, 25, 50 and 100 denominations. The 1 Dalasi note was withdrawn some years ago, and there is now a coin to that value. Unfortunately, the lower value notes are more expensive to print than their face value, so they haven't been replaced recently. Accordingly, they are tatty, often repaired, several times over, and frankly, at times, smell.

Whilst we are there the country announces that it is shortly to replace the currency. Gambia had hoped for a common West African currency unit to be developed, but it has, again, been put on hold, so they are to have a new range of Dalasi notes. The 5, 10 and 25 will be varnished so that they last longer. The thought of a West African regional currency frightens me. A "Euro" in an area of such rapid political change? Perhaps not.

From reception, I go to the pool bar and buy a beer. 35 Dalasis for a Julbrew (the local beer) - not bad. I walk round the hotel. I haven't stayed here before so I need to explore quickly. The rest of the group are still settling into their rooms. The hotel is small and compact. It backs onto the cliffs and we can walk down some steps to a small man-made beach and, at low tide, a bit of natural beach. We have a terrace balcony outside the restaurant which gives a view of Bakau pier and fish-market to the north, and some cliffs and a headland to the south.

Walking back through the restaurant to the bar, I see a lone figure there. He was on the flight on the way out, and is wearing a T-shirt that indicates he is involved with one of the local wildlife trusts. He is drinking the local beer with the stance of a European used to Africa, and I walk across and stand next to him to order another beer. We make polite conservation and discover we both have an interest in reptiles; he is here to photograph poisonous snakes, he has visited previously and his name is Ian.

I tell Ian of our reasons for being here - he is interested and offers to help. We discuss the hotel reptiles; there are some Agamas in full breeding colour, and some five-lined skinks around. I make my excuses and then wander again. Most of the group are emerging from their rooms, my self-indulgent solitude is over. I lead them to the terrace balcony and they admire the view as beers arrive. Lisa has her camera and wants a photo of the whole group for headquarters. I explain about Ian being a photographer, and he arrives as if on cue. I introduce him and he takes the picture.

We decide to eat in that night. Although there are supposed to be some good restaurants in the area it will take time and many are closed because it is the rainy season with few tourists. The restaurant menu is reasonable and the food is actually quite good.

Whilst eating in the restaurant, we discover that there are two cats in the hotel, One is a marmalade cat, and the other is a tabby with some white marks. Both have their ears clipped - a sign to vets around the world that one of their colleagues has already operated, and there will not be any kittens here thank you. The cats ask, politely, for food. Little do we know that these two are to become our friends and confidantes over the next two weeks.

Wednesday 5th July

After a pleasant night's sleep, I am awakened at about 05.30 by Common Bulbuls. *(Pycnonotus barbatus)*. These brown blackbird-sized birds have voices well beyond what could be expected from their size. They sound happy though. The air-conditioning remained on during the night and when I leave the room at 07.00 I get hit by a blast of hot air. Ian is the only one that I recognise at breakfast, so I join him. He is off to Abuko and I tell him that I shall keep the group nearer the hotel today but take them to Abuko tomorrow.

When the group are eventually up and breakfasted we set off for a long walk down to the Kotu area. This interests us for a number of reasons, but mainly because of the alleged burial of "'Gambo'" there in 1983 and because of the silver Ninki Nankas available from the silversmith in the tourist market.

The walk is down the coastal road, and we pass the fire station, the barracks, the MRC and the High Commission. Olly is doing the best that he can, but as a result of the heat, and his weight, he has to take it easy. We rest under some trees for a few minutes and an informally dressed soldier comes across from the elegant compound across the road. Without realising it, we have stopped opposite the house of Jawara. Sir Dawda Karaiba Jawara was the president until deposed in 1994, when he fled the country and lived in exile in the UK. He has been allowed back to Gambia recently, but only on condition that he keeps out of politics. Six odd-looking Europeans stopping outside his house clearly are worth investigating. I talk with the security man about the Gambian army and West Africa generally. He sees the state that we are all in and obviously decides that we are not of interest and goes back to his post. Olly now looks a more normal colour so we go on.

As I am about to cut down from the road to the beach we are joined by a "guide". In fact, I know the way, but he tags along. We reach the beach just off the Fajara Hotel, which appears to be in one of its rebuilding modes. Walking on the beach slightly deters one or two of the local nuisances, but our guide stays with us. As we approach Bungalow Beach, it is easy to see that the new night club, *Destiny's*, now occupies the site by the Bungalow Beach Hotel, and it projects further out onto the beach. Oh dear!

I have the map drawn by Owen Burnham back in my room, but it looks as if digging the foundations for the wall of *Destiny's* may have disturbed whatever it was that Owen buried 23 years previously, if not having actually covered it. We go to the corner of the offending wall and take some pictures. We then walk further along the beach, and enter the tourist market. I am recognised as we approach the silver stall.

The man greets me as I walk towards the stall, ignoring a number of stall-holders who are trying to get me to look at their stalls along the way. I introduce the team, and give the stall-holder a copy of *Fortean Times* 208 which contains my Ninki Nanka article, and the picture of the pendant Ninki Nanka made by his boss. The copy of the magazine is very favourably received and it is shown round. I am not sure if they can read in English, but being *Fortean Times* some of the pictures are jolly interesting too.

They no longer have the pendant-sized Ninki Nankas, but the small ones, in silver, which could be a small pendant or, in pairs, ear-rings, are in stock on Adama Touray's stall and the neighbouring one. Several of the group buy them. The stall-holder will make enquiries, and see if he can get some more of the pendant-sized ones later in the week. As it is quiet, we stay and talk about Ninki Nanka and about other topics. Richard is amazed to discover that one of the men was working there in 1983, when the market was relatively new. What is more, he remembers an animal being washed up on the beach. It was a dolphin, without a dorsal fin. Initially it was alive, and was carried back to the sea, but it vomited and died and so was buried at the top of the beach. The man is certain of this.

I have a problem, because I know a similar thing happened in 2003 when a friend was staying at Kombo Beach. She, too, saw a sick dolphin on the beach that subsequently died. I know, from having stayed there previously, that there are a lot of dead turtles washed up on that beach too. We also found knowledge of Ninki Nanka here. No direct evidence, but one of the stall-holders had an uncle who saw the animal, and subsequently died.

In view of the heat, and differing amounts of energy, we decided to split the group. Three of us took a taxi back, and the other three walked. After a short rest, and a drink at the hotel, I decide to explore again, this time on my own and set off up the road from the hotel, heading north to Cape Point. The locals are annoying, though, with their pestering; they want money, and they want to spend time with you, lulling you into a sense of friendship to make it harder for you to say no, when the request for funding comes. The High Commission resident is up here, next to the Botanical Garden. There are some other high status residences too, and then two high quality tourist hotels. The heat is too much though, so I head back to the hotel, noticing a Black Kite *(Milvus migrans)* flying over the High Commissioners residence.

As I head back to the hotel, I am caught by another "bumster" with the standard "I got married yesterday". I don't break pace, but he is persistent, and eventually asks for money. I really can't be bothered, and against my better judgement give him a £10.00 note. I don't know when he eventually found out that it was 10 "Peckham Pounds" with pictures of *the Only Fools and Horses* cast on it. I hope he tried to pass it on, and got caught.

When I got back, I found that Richard had telephoned Assan and arranged a minibus for Abuko the following day. We swapped stories in the swimming pool, and over supper, and tried to get to bed early. In the early evening, though, Suzi, Olly and I headed off to find an internet café and report home with our findings so far.

When I was in bed that night I discovered that I had got sunburn through my shirt.

Thursday 6th July

A hot and bleary start, I went to breakfast without shaving, not a good example to the other Brits. They used to say that the anti-malarial Paludrine made you absent-minded, and of course there is Paludrine in Malarone.

Lisa drew my attention to a lady at breakfast, I shall call her Jill. They had been talking and it became apparent that she was taking no anti-malarial other than a vitamin B complex. Jill had type 2 diabetes as well! She had been to Gambia before, but not during the rainy season. Fortunately, she was seeing a doctor who, I am sure, would suggest a slightly more efficacious anti-malarial.

Assan turned up at the appropriate time with a minibus and whisked us off to Abuko. I think the drive through Serrekunda surprised some members of the group. The road is a tarmac one, but badly holed in places, and the shops are, well, a bit African. When we arrived I saw that the entrance hadn't changed much in the last few years. Neither had the entrance fee, it was 31½ Dalasis for non-residents. When I was last there that was almost £2.00, now it was 63p.

After a brief chat we walked on through. After a few hundred yards we came to the observation centre, which now has the Darwin Centre underneath it. This overviews the largest of the lakes, which at this time of year is low, very low. The lake was formed when a creek was dammed to form a reservoir many years ago. The lake produced, houses a number of crocodiles, some fish, and a range of amphibians. Accordingly, it also attracts a range of birds. The ones we saw now included hammerkop, black-headed heron *(Ardea melanocephala)* and malachite kingfisher. Lily trotters [jacanas] were also present.

I noticed, at this point, that the group was, unintentionally, picking up one of the guides. They are, allegedly, supplied by the park, and no fee is due. In practice a tip of less than 250 Dalasis at the end of the trip is not looked upon favourably. Some of the guides are good, some are not so good. The not so good ones can be a menace. I wasn't risking it so I snook off. I then went through the next 1500 metres of the park fairly quickly, but quietly, to get to the animal orphanage. Although named the animal orphanage, it is really an area of cages where animals that cannot be released into the park for one reason or another, are kept. There is also a shop where you can buy reasonably priced drinks, and sometimes, snacks or nuts. In fact there was also a woman selling the usual European tourist goods; carvings and cheap jewellery.

The shop was closed and the place quiet so I sat down and drank my own bottled water and watched the (caged) patas monkeys *(Erythrocebus patas)*. A small Nile monitor lizard, maybe two-feet long, wandered past and I took some pictures of it. This seemed to interest the lady selling the gifts, so, when the lizard went back into the woods, I went and talked to her. She told me of a large monitor, maybe six-feet long, which often wandered through the back of the toilets. She then showed me where he (or she) sometimes walked. We continued to talk and I brought up Ninki Nanka. She was aware of it, and modestly scared. I mentioned the Ninki Nanka incident at Abuko in the past, and she stated that the grandson of the man who died was still working here, and his name was Hassan. We talked a bit more and I bought her a Vimto.

A bit later the group arrived, with the guide. They had been spat at by a cobra, which, it seemed, frightened the guide more than our group. Anyway, we all had drinks and the "first-timers" spent some time looking at the hyenas, vultures and baboons. After a time we walked on and, so that Richard and Lisa could buy some books, the guide took us through a back path to the entrance. This caused a bit of a problem as Abuko has a separate exit to the entrance, and Assan was waiting there. Anyway, I went to fetch Assan, and Richard and Lisa re-entered the park to try and find a member of staff who was thought to have a Ninki Nanka scale.

The next two hours were somewhat confused, with park staff trying to find the "scale-owner", and then Richard and Lisa. In the meantime Suzi, Olly, Dr. Chris, and I were stuck in the back of the minibus. Initially, Assan and the park staff, being devout Muslims, went to pray together, after the appropriate washing. After prayers, the park staff received a message that the man with a Ninki Nanka scale had gone next door, to the cattle-market and meat processing plant. It was decided, by no-one in particular, that we should go and fetch him.

Assan drove up there with four members of the team, and a local man, who was well dressed and large. When we got there we stopped outside, what I think was the back of the slaughterhouse, and an exchange took place in the local language. Initially, the man who we came to collect looked a bit bemused and possibly a little scared. He got in the minibus with us, though, and Assan drove back to the entrance to Abuko. It was only later that we found out what had happened. The large local chap who had done the talking, was in fact a police officer, in plain clothes, but who was known by sight to most of the locals as a police officer. He had just told the chap who we went to collect to "get in". A minibus with four whites, a driver whom he had never seen before, and a powerful local policeman suddenly turning up, and ordering you to get in, must have been a little frightening.

Just after we got back to the entrance to Abuko, Richard and Lisa appeared, surprised that they had been away for two hours. They were rather happy that we had two people there to be interviewed. Almost amazingly, both were happy to be interviewed. As both had put themselves out to be there and it was likely to be a few minutes I went off with Assan, and Maryam, who worked there, to buy everyone a soft drink. It just seemed a good idea, so we walked up the road a few yards to a little shop, where we bought 15 assorted bottles of (non-alcoholic) drink. Assan and I carried them back in a crate between the two of us.

When we got back, Richard was interviewing the man about the scale, which he subsequently bought. Sadly, it looked like old celluloid rather than anything of biological origin. Not to buy it, though, would have been a mistake that would have been regretted later.

The interview with Hassan was also informative. He works on the front gate at Abuko, and his grandfather "Papa Jinda" died in the 1940s after two encounters with Ninki Nanka. (One of the travel guides actually puts this death a lot earlier). It appears that the man was a watchman at the waterworks (as Abuko was then). The first encounter did not involve him seeing it, but just involved some damage to pipes there. The second one, a few years later, was to prove fatal, with him dying within two weeks of seeing the animal. He apparently suffered pain in the legs and waist and then his hair fell out before dying.

Bakary Jarju was also present at this time. He came from a village 80 kilometres away, where there was a pool where a Ninki Nanka lived, or lives. Arrangements were made for a visit to the pool on Saturday.

When we got back to the hotel there was much writing up of the day, down-loading of digital pictures on to the laptop and general debriefing on the day. I think we all slept well that night.

Friday 7th July

This was a day of rest for the group, or a day of preparation for the longer trip, depending upon which way you looked at it. We were also still tidying up the report of the Abuko trip for transmission home.

In the spare moments, and there were many with one laptop between six of us, we were able to look at the wildlife in the hotel in more detail. Whilst breakfasting on the balcony I watched a local man, up to his knees in the sea, using a hand-thrown net to catch fish in the sea. I had seen this type of net used in the creeks before, but not in the sea. It seemed to work, but the fish were small and looked like the local, cheap, "*Bonga*".

The hotel revealed a limited, but interesting, lizard fauna; there were agamas in full breeding colours, and five-lined skinks in profusion. Later in the week, Suzi was to discover a gecko in her room too.

As for birds, well a lot were passing through. We had fire finches *(Lagonosticta larvata)* as the resident waxbills, but most birds just flew over. On the sea-ward side there were grey-headed gulls *(Larus cirrocephalus)* and the odd African skimmer *(Rhynchops flavirostris)*.

On a morning, if the tide was out, a reef heron *(Egretta gularis)* would work its way along our bit of beach, always going from south to north, but flying off when it got just north of us rather than risking entering the fish-market area. During the day I saw red-beaked hornbills *(Tockus erythrorhynchus*), some ringneck parakeets *(Tacula krameri)*, a long-tailed shrike *(Corvinella melanoleuca)* and several speckled pigeons *(Columba guinea)*.

Saturday 8th July

Richard, Dr. Chris, Suzi and Olly head off with Assan, and the chap from Abuko, to Dumbutu. Lisa and I have decided to stay at the hotel so that we can make a more detailed study of our respective bathrooms. The chamber maid guesses what is wrong with me and asks if I have any diarrhoea medicine for her? I don't, I do not know her medical history, she could be pregnant, and some of the medicine I have with me is contra-indicated in pregnancy, besides at this point I feel I might need the whole packet myself!

Now I know that I am back in Africa!

Whilst I sat on my veranda, I see my first sunbird of the trip, a splendid sunbird *(Nectarinia coccinigaster)* - it feeds from the hibiscus, but disappears before I can point it out to anybody. The five-lined skinks also provide much entertainment. There are some babies around as well. The little patch of garden in front of our block of rooms, maybe five-feet by fifty-feet, seems to house at least three adult skinks and four or five babies. They climb as well; I saw one go up the palm. I spend most of the afternoon trying to get a decent photograph of them, but because of their size and tendency never to stay still long, this is actually quite difficult.

The team return safely in the evening with much to talk about. They haven't seen a Ninki Nanka, but they have had another African adventure.

Sunday 9th July

Although feeling better, I still need to watch fluid levels and not go too far from the room. Ginger the cat comes round and visits the sick and ailing. He seems to take this job seriously. I sit on the veranda and try to develop a storyline for a new novel, and Ginger sat next to me, more like a loyal dog. Suzi comes round and asks how I feel; as I am feeling the heat I decide to go and lay down in my (much cooler, air-conditioned) room.

I turn to leave her, opening the door. I look for "Ginge" as an afterthought, but he is not there. Semi-laying on the bed I feel better already and turn to lay on my side and write in my journal. After a few minutes I feel gentle pressure on the back of my neck. For a second, I wonder whether I have let myself slide into dehydration and salt imbalance, producing cramps, but I have never had cramp in the back of the neck before. I turned round to get more comfortable, and nearly had a heart attack. There on the bed, having arrived silently, is "Ginge", just sat looking at me and occasionally leaning forward just to touch me. He had sneaked, quickly and quietly into the room as soon as the door was open.

I tell him that he is a good cat, and ask him, politely, to leave. He does so gracefully. Later in the day, feeling refreshed, I go on the evening walk to the internet café to send off the daily blog. On my way back I pop into the supermarket and buy a sachet of Kit-e-kat. "Ginge" pops round that night, after restaurant duty, as I am having a drink with Suzi on her veranda, and we give him a late supper. He clearly appreciates it, and as soon as he has finished he starts rubbing himself against one of the plastic water bottles. We pour him water and he drinks, thirstily, for several minutes. He wanders off for a few minutes and then comes back and listens to us, now the whole group, as we talk of Africa.

Monday 10th July

I am feeling a lot better and now cut down on the relatively powerful opiate that I was taking. An early morning walk, well 9.00am is interestingly still too early for bumsters, took me up the road to the movie café. Olly has seen a poster advertising a DVD set of *Banjul Cops* [1], the Gambian home produced edition of their equivalent of *The Bill* or the *Inspector Linley Files*. Olly is desperate that we get a copy of it before returning home. What he hadn't noticed is that the poster is outside a building labelled "The Movie Café", and a bit further up, "Vinasha Productions". I realised that this is actually the production house of the series and go in.

There is a neat entrance with a reception desk. The receptionist is attractive by both African and European standards. I ask her about the series, and she hands me an advertising sheet. The boxed set of DVDs is available for 1500 Dalasis. This is thirty pounds, a little bit more than I was intending to spend, but I may never come here again so why not? Besides I have friends who work in TV in the UK who would love to see it. I purchase a set. It takes ages because I have to have a written receipt, and then it has to be wrapped. By the time I return to the hotel there is the classic shout of *"Excuse me"* from the bumsters sat opposite the hotel.

We are to go to Banjul today. The trip is multi-purpose. To see the museum, get some music and visit the Atlantic Hotel. Assan arrives at 1.00pm and recruits another

---

1. **Banjul Cops,** The Gambia's ground breaking 10-part police series produced in 2003, has already been syndicated to a number of broadcasters in Africa including M-Net and Pulse Africa TV to critical and popular acclaim and has had rave reviews from the UK based Black Film festival.

taxi driver locally so that we travel there as a two-car convoy. We park at the Atlantic, and that is to be the rendezvous point in a couple of hours if anyone gets lost. Declining the offers of a guide, I lead the way into town past the Royal Victoria Hospital, the School of Nursing, and another building that has been taken over by the state security forces. We come along past the museum and what used to be McCarthy Square, but is now July 22nd Square, named after the day when Jawara took over. Our first port of call is to be the Methodist bookshop, but looking for a cashpoint as well. I try my best, but cannot find the bookshop. I see the police station so we all traipse in there and I ask the desk officer. He says he thinks it is still there, within two blocks, but he just cannot remember where. A plain-clothes officer comes outside with us to show us the way. The only problem is that he doesn't know where it is either. Dr. Chris does see a cashpoint though, opposite the police station, so we go back for that. The only problem is that I am the only one who can get any cash. Going into the bank doesn't help much. There are queues everywhere! We move back towards Albert Market.

I go to the market to try and get a copy of *Ninki Nanka* by *Toure Kunde*, it is a classic of African pop music. Unfortunately, you just cannot get it anywhere in Europe now. The stall on the outside of the market does not have it, and our self-appointed guide, "Don't worry I am a security man" – he shows me a photocopied I.D. which frankly could be for anyone of any race, except for the name that is - knows of another stall. I follow on a circuitous walk, where we go upstairs in the long building and move through a number of women making clothes. I have instructed the rest of the group to stay behind. They form a close group at the bottom of the stairs that I have ascended, by now with three locals.

Eventually, after climbing over several seamstresses and their work, we come to a stall which is clearly locked up. One of my guides says, *"He gone tinkle, back soon"*. The amount of dust on the padlock might suggest otherwise, so I say there are other stalls. The lads agree and we go back down some stairs to the main market at street level, with me leading the way. Miraculously, we come down the stairs opposite a stall that is manned and that has the title I want. I sit on a stool whilst the cassette is copied, (ironically under a sign which says *"illegal copying is theft"*). Whilst Gambia supports the initiative against copyright theft, all the music stalls in the market have CD and cassette copying facilities.

Having got the cassette, and paid a slightly over the odds price for it, I make my way back to the group, not by the long route back through the market that my "guide" wants, but by going outside onto the street and walking along the side of the market before re-entering the market by the stairs where they are standing. They are all looking up the steps that I disappeared up and are apparently a little concerned. It confuses them that I came back at ground level and from a different direction. My guide is about to discuss his fee for guiding me to such a rare piece of music. In fact, he hasn't done anything that I couldn't have done myself. So I say "Right come on troops we are out of here" and set off back towards McCarthy, oops I mean 22nd July, Square.

The guide wants *a "small, small fee",* maybe 200 Dalasi. Without breaking pace I give him 25 and say he hasn't done anything, he has wasted my time, not helped me. The 25 Dalasi is apparently an insult for a security man, but he holds on to it. We are moving into more public areas and he can see that he is not going to get anything more out of me, so he moves off. By the time we get to the square we are on our own again, despite the square being crowded. After a quick conference, we decide to give up on the Methodist bookshop as it seems to have gone; well if the police can't find it, it must have, so it is on to the museum.

At the museum it is 50 Dalasi to enter, and you must deposit your bags and cameras at the entrance. I am parched and so I pay for the group and then head to the outside kiosk. Two of the group are complaining that it is £1.00 in UK currency to get in here when it is only 63p for Abuko. I point out that the Cokes are only 20p and that there isn't anything I know of in the UK that you can still get into for £1.00. This seems to calm the issue.

The museum hasn't changed since I was there last; a bit more dust maybe, but no other changes. I had a quick look round and found nothing of cryptozoological interest. The only thing I did find that even vaguely interested me for the purpose of this mission was two pictures of Captain C H Armitage, the governor of what was then the colony of The Gambia from 1920 – 27. Predictably, he was with various dignitaries and not with his skink!

After some time at the museum we went on to the Atlantic Hotel. Acting as if we were residents, I walked straight through reception and out to the pool area. The idea was fairly straightforward. I was going to order six Chapman's, and we would drink to the memory of John Downes senior. We were, after all, on the J. T. Downes Commemorative Expedition.

John Tweddell Downes was CFZ Director Jonathan Downes' late father. He had spent some of his working life in Nigeria and was very fond of West Africa. In his last few months he had discussed the possibility of this expedition with Jon and Richard, and on one occasion with me. As some of the collection at the funeral was going to finance this trip, it seemed only right that we should have a moment to commemorate his life, and with the Atlantic being one of the places with a long colonial service involvement (the old Atlantic Hotel preceded the tourist industry in Gambia, and often housed newly- arrived colonial employees) it seemed the right place to do it.

A Chapman's is a non-alcoholic mixed fruit drink which is only ever made properly in the Gambia, and the Atlantic Hotel is its home. Well, it used to be. I know the hotel was quiet, but I still do not think that the explanation that they could not make Chapman's because they had no cocktail shaker was acceptable, apart from anything else it can be stirred and not shaken! Anyway, we weren't going to get one, let alone six, so I ordered six fruit cocktails instead and we drank to Mr. Downes' memory with those. After a few minutes of sombre reflection, Suzi saw a male agama that was showing off on a palm tree and went to try to photograph it. The agama was obviously a little shy, and it kept on moving so that it was just out of position for a good photograph.

One of the points of visiting the Atlantic was to see the "bird garden". This is a square with some fairly mature trees in it and a raised viewing platform. It is positioned between the new Atlantic (completed 1980) and one of the blocks of the old Atlantic Hotel. Although only a small square of trees, maybe two hundred feet by two hundred feet it is an interesting block of trees, and there is, additionally, a path system through it at ground level. It forms a home, or at least a regular stop-off point for a number of birds, and also some five-lined skinks and a colony of Gambian epauletted fruit bats. Whilst we were there, the birds were really limited to several pigeons and doves, but the fruit bats were present in great numbers and, although not really flying, they were certainly doing a bit of pushing and shoving amongst themselves. Unfortunately, whilst watching the fruit bats we were all attacked, at about the same time by a swarm of flying insects that seemed to take you between ground level and two-feet or so above ground level. They settled on you in swarms and bit. They were clearly after blood, but at least had the decency not to follow us out of the woods.

After leaving the bird garden we had a quick look at the hotel shop and a couple of stalls in the hotel garden and then went back to our hotel at Bakau. One of Lisa's pictures of the fruit bats went in the blog, and the UK team inserted a red circle around one of them just to confirm where they *really* were in the picture.

When I shower in the evening, I find a large ant head embedded in my leg. I guess that got me at the same time as the biting fly attack.

Tuesday 11th July

Olly's birthday

Another day of rest and preparation, and I'm on the sick list again! This time it is mosquitoes! During the night I have been "done", and more annoyingly it appears that I am allergic to the bites. I go up the road to a local pharmacy and buy a medium strength steroid cream. The pharmacist is a chap from India and we talk medicine for a bit. The steroid costs the equivalent of 40p; if I wanted it in the UK I would have had to pay £6.25 prescription charge for it, as it is a prescription only medicine here. I also would not have been served instantly or engaged in an interesting conversation for 10 minutes.

Later in the day, we go to the internet café and send yesterday's blog and photos off to base camp (Woolsery). I get cat food at the supermarket again. Feeding the cats is becoming a habit, although there is no real evidence that they are underfed. "Ginge", in particular, almost seems to say thank you for his food. I don't think I have met as good a mannered cat before! On the human food front, I found some banana and cinnamon biscuits in the supermarket which I took back, and they proved very popular with the evening drinks.

Wednesday 12th July

"Starling Revelations at Treason Trial" (sic) was the headline on today's *Daily Observer*.

It rained during the night - seriously rained, as it does in the tropics during the rainy-season. The weather this morning feels less oppressive as a result. The improvement in the atmosphere, though, comes at a price and there are now a lot of mosquitoes on the room screens.

Richard organises a trip to Kartong to the sacred crocodile pool, the Allahein River and the "reptile park". Although there is a good road down there, I'm still not sure about the trip and back out.

I decided not to go on this trip, as I had a number of things that I wanted to look at locally. In fact, it was the right decision, because it enabled me to record a local, short-term, pollution incident. When I walked down to look at the sea from the lower hotel terrace at about 11.00 am, there was a line of fish and fish guts along the full length of the hotel beach. All the fish were twelve inches or less in length, and they were the local ones that were the cheapest in the fish-market. The guts, were, well guts, and I couldn't work out what they belonged to, except that whatever species it was, they were from individuals that were bigger than all the dead fish.

I walked further up the coast and then back down to a point just South of the hotel, and found that the strip of dead fish stretched maybe half a mile. There were several thousand fish involved. At one point I found a month old newspaper in with them. The newspaper was in Spanish. It would be improper to say that this tide of death was from one of the European trawlers, trawling twelve miles plus offshore, but the circumstantial evidence did suggest that it was, and the sheer volume of fish suggested that it was too big an incident to have been produced by the local fishing boats. Many of the dead fish were of a size and type that could have been sold in the market anyway. One of the hotel staff suggested that these tides of dead fish occur every couple of weeks.

The group are back in the evening, having had a wonderful time. They have seen and photographed a baby crocodile at the pool at Folonko (Kartong), they have visited the reptile park and got information on Armitage's skink, and they have been in canoes on the Allahein River and crossed the border (just) into Senegal. They all feel like real explorers now. Additionally, when Dr. Chris types up the report I discover that there is a claim that there is a white crocodile at Folonko. It is said that if you see it you will become a great person.

Thursday 13th July

At 2.00 – 3.00 am there is a real storm that wakes us all up, heavy rain, thunder, not much lightning though, and power cuts. The hotel has its own generator near the entrance which has been operating occasionally since we have been here. It now goes on for most of the day. The Kartong trip was typed up and we attempted to email it home. The internet cafés are not open for business though. Although some of them have generators, they will not risk the computers on these generators, presumably in case of voltage surge. Ironically, our café has a voltage regulator that they use on the mains (when it works), so it should have been safe.

Following a text from England, Richard is on stand-by for the BBC World Service (Africa) to call for an interview. They do, before lunch, and he gives the interview. He has to move into the office, though, because of the noise in reception. The power continues to go on and off.

Whilst at the snake farm yesterday Richard found out that Luc, the owner, had two Armitage's skinks that were caught by the café owner at Gunjur. Gunjur is about 35 kilometres south of us. If Armitage's skink are there that adds some distance to the known range. The first ones were caught one kilometre to our north, so this is good news for the species. Unfortunately, as the whole coast is being developed, and as far as we know these animals only exist in the coastal sand dunes, they are still in potential danger. It is something that we should investigate.

Richard telephones Assan for a quote for the trip and then finds that the local taxi boys will do it a lot cheaper. We decide on one taxi, and I have an idea for a short investigation at Kotu, so I say that if Olly wants to come fishing at Kotu he can come with me. Olly, though, changes his mind, and they end up taking two taxis and the group sets off. I have other plans.

I do a little bit of exploring in the Bakau area, and manage to find a local internet access that was working. From here I am able to print off the article about Richard, and the trip, that was published in *The Independent* whilst we have been here. I am also able to collect a silver bangle that I have had made by a local jeweller for Lady Jane.

When I get back to the hotel, reception tell me that Hassan from the BBC has called and the interview has been broadcast and the story is already up on the website. The hotel seem impressed that the BBC have actually telephoned them! The gardener calls me Ninki Nanka man as I sit and have a drink.

When the group return, they have had a disastrous trip. They have found the place and spoken to the chap who caught the skinks - he runs a café. However, they attempted to observe dunes were it was alleged to live, and get nowhere through being plagued by a man and his dog who want to help. They try and get rid of him but he just takes the dog home and reappears.

Friday 14th July

Lisa's birthday, and we decide to get back to the `Gambo` story. It's taxis for 11.00

and we plan on using Assan for two trips. He, in turn, recruits one of the local lads so we travel in convoy again.

Before we leave, though, we speak to Babouka, the front house manager at the hotel. He knows of a local incident about Ninki Nanka. In the 1960s, maybe about 1965, there was a large animal seen entering the sea at Fajara (one kilometre south of us). It was seen by Bishop Maloney and left a trail. A warning to stay clear of the area was apparently given on *Radio Gambia*. Typical, you decide to spend a day on `Gambo`, and a good Ninki Nanka story comes up, and one that could involve investigations both locally and investigations in Europe (Bishop Maloney, did he keep a journal?).

We get our taxis to drop us at the beach below Palma Rima and start to walk up the beach towards Kololi Point. I am still looking for fish remains from the pollution incident. I find a few that have the right degree of decomposition. Things disappear fast here with the heat and the carrion feeders. (Two vultures decide to leave South Africa, they have had enough of the violence there. They go to the airport to get on an SAA flight to London, one has a dead impala fawn under its wing. The girl at check in says *"do you want me to tag that so that it can go in the hold?"* looking at the fawn. The vulture looks her straight back and says *"No its carry-on."*)

The beach bums become a nuisance so we turn and return. Palma Rima is closed, but there is an internet café attached to an establishment called Luigi's that is open. It is run by a chap from Reading, whose accent, for some reason, I mistake for South African. We use some machines and get up to speed with news from home. Suzi is able to send off the blog for the CFZ at home. The chap puts BBC news on satellite TV up for us and we decide to stay and eat here. Everyone has pasta except for me. I have an omelette, and it is good.

There is a story, that I think started with Assan, years ago, that before the Palma Rima was built a Ninki Nanka lived in a hole nearby, and whilst it was away the locals planted a tree in its hole to stop it returning. One of our missions today is to find that tree! Because of its sacred nature, I guess the tree would be either a baobab or a silk cotton.

We walk up to western wall of Palma Rima, which is closed, and then take the bicycle track, now paved, north towards Kotu. This brings us to an area of ground where there is scrub and rice paddy. We have the main road to our right and the sea to our left, both a few hundred metres away. There are a number of groups of trees here. There is a big baobab that I have a feeling about and we approach it. There are multiple carvings and indentations on it. Additionally, there are holes that have been made in the trunk which contain cloth, and others that contain paper. The tree certainly has some significance.

Suzi, quite rightly, points out that this tree is much older than the Palma Rima Hotel, and that if the tree was planted to stop Ninki Nanka returning just before the Palma Rima was built, then this tree must be too big, (Palma Rima first opened in 1990). I agree, but there is something about this tree…

We look at others, there is a smaller baobab nearer the road, but it is too close to the road and that road has been there a long time. Too many trees and no local guide who was there at the time. One of the local birdwatcher guides comes along, a qualified one; he has smart clothing and his badges of office, binoculars around his neck and a bird guide in his bag. Do we want a guide for birdwatching? No we don't, does he know anything about Ninki Nanka? He has heard of us, he thinks maybe we should look up river for Ninki Nanka. He knows of nothing local.

We move on up the bicycle track, having given up on trees. Suzi asks me about the plastic bottles in the tops of the palm trees, and I explain that they are there to gather the juice which is later fermented into palm wine [1]. As we near the Badala Park Hotel we are looking more and more at the rice paddy on our seaward side. Lisa sees a mammal run along the edge of one of the paddy mud boundaries. Thinking that it is a mongoose I follow through. In fact I get one quick flash of it as it disappears into the greenery and it is a pouched rat *(Cricetomys gambianus)*. They come big in Gambia!

The birds here are impressive too, a great white egret *(Casmerodius albus)* that is close enough to photograph, lily trotters and a group of white-faced tree ducks *(Dendrocygna viduata)*. Richard sees a sacred ibis *(Threskiornis aethiopicus)* as it flies over and a squacco heron *(Ardeola ralloides)* also makes an appearance. We move on and join the road as it turns down to Kotu Strand and the hotels. As we pass the end of Kombo Beach, there is a flock of fire finches which we stop to photograph. There are cordon bleus there too and I point them out to Suzi and Lisa. We move on to the tourist market and make some more purchases, before going through Bungalow Beach to the area by Destiny's where "`Gambo`" is allegedly buried. There follows much excavation and much attention to what we are doing, but we just go for it. Our equipment is not too high tech, a garden trowel (plastic from Mister Pound) and two children's plastic beach spades (small).

We have acquired a couple of locals, and we explain that we are assessing beach erosion. Dr. Chris has a hand-held GPS unit which can be made to bleep, so he claims that it is some type of recording device. A third local lad joins us and we explain that we are using this point, as the wall on Bungalow Beach has been there since about 1980 and hasn't moved, so it is a reference point. In fact, the cracks in it suggest that it might be a centimetre or two closer to the sea than when it was first constructed.
A local fruit juice vendor comes and tries to sell us juice. He has a lot of local man-

---

1. **Palm wine**, also called **palm toddy** or simply **toddy**, is an alcoholic beverage created from the sap of various species of palm tree. The drink is particularly common in parts of Africa, South India (particularly Andhra Pradesh, Kerala and Tamil Nadu, where it is known by the name of **kallu** in Telugu) and in the Philippines, where it is known as **tuba**.

The sap is collected by a tapper, who cuts between the kernels of the tree. Some sort of container, such as a gourd or plastic bottle, is left to collect the draining sap for a day or two. The initial white liquid that is collected tends to be very sweet and is not alcoholic. The sweet white liquid before fermentation is called "neera" and is refrigerated, stored and distributed by semi government agencies in Maharashtra. Neera has a lot of nutrients including potash. However, the sap begins fermenting immediately after collection due to natural microorganisms in the air (this is often spurred by residual yeast left in the collecting container). Within two hours, fermentation yields an aromatic wine of up to 4% alcohol content, mildly intoxicating and sweet. The wine may be allowed to ferment longer, up to a day, to yield a stronger, more sour and acidic taste, which some people prefer. Longer fermentation produces vinegar instead of stronger wine.

goes that he is clearly wanting to juice, and to get rid of him I say that we will have a juice each when we are finished in about 40 minutes. The security people from *Destiny's* come to see why we are digging up their wall. We explain the beach erosion mission, and start asking them about the construction. They don't know, but seem happy with our explanation. In fact, we really are starting to get a feel for the dynamics of this beach. In the first hole we find what looks like virgin sand at about 50 cm down (this is difficult to assess, but there is a clear horizon exposed where the sand changes colour). Although the tide is out, the slope on this beach is gentle and the depth that we are at is one at which the water will get to at high tide.

We try a second hole further down, still nothing. Our third hole would involve digging up a young, but sturdy, fan palm and it might be pushing our luck a bit. We decide to call it a day. I move down the beach with the local lads, who now all want to be scientists. This gives the others a chance to have a straight conversation and fill the holes in without interruption. I dig some sand out near the sea and show them the speed at which the water refills and the way in which the sand changes colour as it drains. There is in fact some dark organic matter in the sand, which seems to come from offshore, but it lightens off and disappears after a day or two in the sun. The beach is a famous and beautiful one, but the sand is not quite perfect. As I am telling them this, a chap comes up and says *"I know where you can see Ninki Nanka"*. Clearly, our alternative identities have not hidden our original dragon-hunter identity. The man with the Ninki Nanka knowledge, is clearly trying it on though.

At the top of the beach, the juice man is talking to the *Destiny* security man who has walked across to him. Through the limited amount I can hear, the juice man is saying something about trying to find out if the club is safe! Oh dear, what have we started!

The mango juice was excellent and we went back to our hotel by taxi. There was a lot of sand left in our showers that night.

Saturday 15th July
(or Sunday 14/7/06 as the hotel notice board claims)

The barracks across the road seem to becoming more secure. During the week, the guard room has acquired a sandbag emplacement, the guard on the gate is now armed with a Kalishnikov, and there is a manned machine gun post covering the gate. The arc of fire on the machine gun covers the road, and the south side of our hotel across the road. There is a lot of military activity in the street too. There are treason trials going on at the moment in relation to an attempted coup back in March. [1] As these involve several army officers there may be things happening that the public are not aware of.

I walk to the supermarket and try to get a paper. There aren't any. I buy cat food and more banana and cinnamon biscuits. There is raspberry tea, in tins, so I buy a tin for Olly, because he loves his teas. When I get back to the hotel, Richard has had a telephone call from Maryam at Abuko. She has found a man who knows where there is a

Ninki Nanka living currently, in a hole not far from the village of Lamin, which is not far away. It sounded good, if it checked out we would go tomorrow and pick him up at Abuko on the way past.

A later 'phone call, though, proved to be a disaster. The man wanted us to pay £5,000 and it was pounds, not Dalasis. He thought that this was a fair price as he could die if he came with us. Also, we would have to take a live dog to throw down the hole in which Ninki Nanka lived. At this stage of the trip we would be lucky to get £500 together between us, let alone £5,000. There would be no guarantees, and we were certainly not going to sacrifice a dog, or any other living animal. Richard was firmly of a belief that the man was a con-artist and he immediately declined. He did, though, get the approximate area, which was the mangroves near Mandinari.

Richard is a dog lover, and not particularly fond of cats. That night he tried to discuss with "Ginge" whether he would like to come out with us tomorrow and look down some holes. We think he was joking, and the rest of the group would probably have put him down any suspicious hole first rather than risk the cat, who has become a good friend and confidante.

As we were finishing dinner tonight, Lisa shouted and pointed. Outside, on the terrace, a Gambian pouched rat walked past. A couple of us nipped out to see if we could watch it for a bit, but it was gone.

Sunday 16th July

The mosquitoes have visited me again in the night and I have a bite on the eyelid which has swollen and closed my left eye. Not really seeing what I am doing, I spray the bathroom for flies using my (locally purchased) deodorant instead of the fly spray. With the fly spray they fall a few minutes later and buzz a bit then die. With the deodorant they just fell and died! A little worrying!
We are off to Mandinari creek today to see some real mangrove, and then on to Abuko. Mandinari is a large area of mangrove and is where our "get rich quick man"

---

1. From *Wikipedia*:

Following the coup in July 1994, a presidential election took place in September 1996, in which retired Col. Yahya A.J.J. Jammeh won 56% of the vote. Four registered opposition parties participated in the October 18, 2001, presidential election, which the incumbent, President Jammeh, won with almost 53% of the votes. The APRC maintained its strong majority in the National Assembly in legislative elections held in January 2002, particularly after the main opposition United Democratic Party (UDP) boycotted the legislative elections. On the 21st and 22nd of March 2006, amid tensions preceding the 2006 presidential elections, an alleged planned military coup was uncovered. President Yahya Jammeh was forced to return from a trip to Mauritania, many suspected army officials were arrested, and prominent army officials, including the army chief of staff, fled the country. There are claims circulating that this whole event was fabricated by the President incumbent for his own devious purposes - however the veracity of these claims is not known, as no corroborating evidence has as yet been brought forward. It is doubtful whether the full truth will ever be known however, as anyone with any evidence would not be likely to come forward with it in light of the poor human rights record of the National Intelligence Agency, and their well-known penchant for torturing and detaining indefinitely anyone who speaks up against the Government.

We would like to stress, however, that this excerpt is included here for compolteness sake, and the CFZ are not taking any political stand in this matter.

suggested he could guide us to a hole where Ninki Nanka lived for £5,000.00 plus one (expendable) dog. At the last minute, Dr. Chris decides that he cannot come; "bathroom study leave", as I now call it.

Assan has borrowed a minibus again and we drive out through Serrekunda on to Abuko and past. A few miles past Abuko at Lamin we take a right and go straight on to the most appalling laterite [1] track. After two hundred meters, though, this becomes a good road again - quite strange. We carry on to the village at Mandinari, and Assan asks directions to the "beach". I have been here once before in 1993, but cannot remember the way. After driving over a rough track we get eventually to somewhere that I recognise. There is a palm-built jetty that runs out through the mangrove to a river, it is not the main Gambia River, but one of the deeper creeks off it. To our left is an area of mangrove which links up with the mangrove behind Banjul, 8 kilometres away. This is a big area of mangrove with many creeks running through it. The Mandinari creek actually joins the Lamin creek a few kilometres from here, before it, in turn, joins the Gambia River.

If there was a Ninki Nanka in Abuko, through which the stream that joins Lamin Creek runs, it could move along the water course to this area, and vice versa. As there are about fifteen square kilometres of mangrove here, and it is taller than we are, we are not going to enter it. It would be too easy to get lost. There is, potentially, quicksand and the river is, of course, salty and tidal here too. As we talk to the local children they say that there are crocodiles here. There are not supposed to be, the river is far too salty. The Nile crocodile has been found on beaches in Kenya, though, and I know that the crocodiles from the sacred pool at Bakau do sometimes end up in a salty creek a mile or so away from their freshwater pool.

I test the children by asking about hippos here. No, they aren't any. If I want to see hippos ("Ippopos") I have to go to middle river division. That is in line with received wisdom. Perhaps there are crocs here. There are streams running into the Gambia River so they may have a much higher freshwater content than the river itself. The jetty is long, and has been repaired since 1993. At the end, there is an open creek, which must be quite deep because there are children running on the platform at the end and jumping into the creek to swim.

Most of the group look at it suspiciously. I say "stay six feet apart at least" and lead the way. Suzi moves in behind me, then, I think Lisa. I am too busy keeping my eye on which strips of palm "wood" I am actually treading on. I'm about fifteen metres along when I hear a shout go up behind me, *"Get off Oll."* It is Lisa's voice. Lisa had

---

Laterite is a surface formation in tropical areas which is enriched in iron and aluminium and develops by intensive and long lasting weathering of the underlying parent rock. Nearly all kinds of rocks can be deeply decomposed by the action of high rainfall and elevated temperatures. The percolating rain water causes dissolution of primary rock minerals and decrease of easily soluble elements as sodium, potassium, calcium, magnesium and silicon. This gives rise to a residual concentration of more insoluble elements predominantly iron and aluminium.

Laterites consist mainly of the minerals kaolinite, goethite, hematite and gibbsite which form in the course of weathering. Moreover, many laterites contain quartz as relatively stable relic mineral from the parent rock. The iron oxides goethite and hematite cause the red-brown color of laterites.

apparently climbed on to the jetty after Suzi, and then to her horror, Olly had got on it behind her. Lisa was clearly scared that the whole thing would collapse. It really felt to be a frail structure and many of the "treads" were bending downward a good two centimetres as I walked on them. Olly, well Olly is a big lad, and he must weigh considerably more than me.

Olly got the message; apparently he was only trying the first few yards anyway, but it was dangerous. He frightened Lisa that much that she dare not go any further. Suzi and I continued to the end, whilst the rest of the group walked along the edge of the mangroves, looking at fiddler crabs and mudskippers. When we got to the end of the jetty, Suzi and I stand and talk to the older children for a few minutes and then head back. It is a relief to be off that jetty; falling through it could be a very unpleasant experience.

We walk round the edge of the mangrove and meet up with Lisa, Richard and Olly who have seen lots of fiddler crabs, some birds and a few mudskippers. We decide to leave the area. The group now have some idea of the width of the mangrove before you get to the river, and the fact that something big could be there without anybody locally really having any idea about it.

As we head back towards the main road, Lisa explains to Assan that she wants to pick a mango from a tree. Assan sees a suitable tree and stops. Everyone piles out, Assan points out a suitable mango, and Lisa picks it, whilst Suzi takes a photograph of her doing so. We get back in and head back along the main road to Abuko.

When we get to Abuko, we are greeted by the gate staff who knew that we were coming. Maryam and Richard discuss the man who she had put in touch with Richard who wanted to charge us so much. £5,000 in Gambia is a fortune, and Maryam is disgusted with him. We talk for a bit and then pay our admission and enter the park. This time we end up with a guide attaching himself to us who is hopeless. I see my first orange-flanked skink and he scares it away! Still, on our way round we do see a number of things that we haven't before, including eleven yearling baby crocodiles, ironically in the pool in which they try to segregate the big crocodiles. I see a violet plaintain-eater *(Musophaga violacea)* for the first time in Abuko. There is a profusion of other birds too, although none as colourful as the violet plantain eater.

When we get to the animal orphanage there are two European couples there with guides. The two men look like professional footballers and their partners are "trophy wives", with perfect make-up and not a hair out of place. In this climate, being a trophy wife so perfectly presented must be hard work. The chaps have a certain air of superiority which is not common in young men of that age, and older people might say comes from too much money when you are too young. It is only an air though, if they are famous we don't know who they are, and frankly we probably wouldn't care. The leveller comes from one of their guides, whom we have not met before. He looks at us, then looks at us again, and announces, in a loud voice, *"Ah the Ninki Nanka Safari". We* are the famous ones here today as far as the locals are concerned. Our self-appointed guide actually starts to lead the group the wrong way when we near the exit. I get them back on the exit path and we have a close encounter with one of the western red colobus monkey groups, getting closer than we have done before. As we leave, he is asking for money and we each refuse. He is furious, and we are too: he has been a nuisance and we made it clear to him from the start that we do not want him with us.

Outside, things are different. Assan is across the road with Max. Max is a Rasta man with his own little road-side café which serves soft drinks from a cold box. I remember Max from one of my college trips, and although some of the others want to go back, I suggest that we have a drink. It is a good decision, we talk about things, about my students, one of whom was also Rastafarian. Max finds out that Olly is from Wales and shows him a Welsh flag and shouts *"Iechyd da"* at him. Olly is elated.

We discuss Ninki Nanka with Max, he hasn't seen one himself, but as a boy he nearly drowned in a river which was rumoured to have one in it. He was pulled out and successfully resuscitated. Since then he will not bathe in the river. He told us where this was and then went on to say that Eddie Brewer, the man who established Abuko, had imported big mirrors to scare the Ninki Nanka away from the pool in there. This was interesting, and we hadn't heard it before. As the Brewers only came to the Gambia in 1957, and later became interested in Abuko, this would put it well after the death of Papa Jinda. Further investigation on this one could be difficult though. We left Marx with many hand-shakes and good wishes for the future.

When we get back to the hotel we say our good-byes to Assan. He says that if he is free he might come and see us off on Tuesday. I say that if he has the offer of any work he mustn't.

Tonight we relax and assess what we have achieved, and what we haven't.

Monday 17th July

We put our final thoughts together today, and get them on to Suzi's laptop before heading up to the internet café for the last time to get them off. Four of us go to the *Daily Observer* office first though. It is a matter of courtesy. If they want the story they can have it. If not it is a matter of editorial judgement for them. We take the memory stick with us, it has all the blogs on if they want them as a basis for the story. The Observer is a ten minute walk up the road, past the internet café that we have been using. We walk up there purposefully and are stopped at the entrance by an older man. He appears to be a security man of sorts. I give our story. He decides that we are vaguely worthy of admittance and takes us up the stairs at the side of the building.

Lisa is concerned as she thinks the offices must be downstairs. The upstairs of a national newspaper office would seem an unlikely place for four of us to get mugged at once, so I'm not too worried. No need to be, we go into the office of the editor and there is chaos whilst they get the chairs for us. We tell the editor what we have done.

He isn't too interested but sends a reporter in. The reporter is a youngster who turns out to be the agricultural correspondent. I write names down for him and we give him our story. After about ten minutes an older man comes in and interrupts. He is apparently one of the senior reporters. Are we the people who came to search for Ninki Nanka? Have we heard the story on the BBC World Service? We all point at Richard. I say *"He was the man interviewed"*. The senior reporter looks at our agricultural correspondent and says "this is an important story, can you get the website?"

We are not sure which website they want, certainly we can connect them to the CFZ, which, anyway will have more facts than the BBC one. So we say that we can get it. The young reporter was then given a memory stick, and we all head off to the internet café four doors up the road. It appears that the *Daily Observer* does not have internet access! Further chaos then ensues - although we can get the website we can't save it on that machine so we end up copying our memory stick on to their memory stick, but not until Dr. Chris has cleared enough space on their memory stick by deleting stories on it already. As we are about to leave, the reporter asks *"who is going to pay?"*

It was only 20 Dalasis (40p) but I end up paying again! The newspaper apparently does not have 40p for a story. We go back to the newspaper office and they want a picture. We sit down again, in an outer office where there are two computers, at one of which there is a girl asleep, absolutely out cold, and everyone ignores her. Indeed, they make efforts not to wake her. I take a second look, and actually check for respiratory movements, as sick jokes about people dying at work go through my head. It is alright, she is alive.

The photographer comes out of an office. He has a little digital camera. He takes us outside, onto the upstairs terrace where the senior newspaper people have their dining table. We are stood against a mango tree and two pictures are taken. Despite it being a bright day, the camera flashes, and makes my eyes water, it is so bright. I tell our reporter that Lisa loves mangoes, and he picks her one from the tree behind. He says that it is a good one. I suppose as the agriculture correspondent he should know. We leave, and go back to the hotel.

That night we have a special dinner, and discuss the low spots and high spots of the week. We go back to the rooms, and sit and talk whilst drinking. "Ginge" and BB come round for a late supper whilst we talk. I shall miss those cats!

We discuss our highs and lows of the week. I describe seeing a mosquito standing in a puddle of deet (insect repellent) on my abdomen and biting me as I lay in bed as a low. My high related to the group as a whole getting so much Ninki Nanka data.

Tuesday 18th July

Our last day in Gambia, and really it is only half a day. I wake early and have a simple breakfast on my own. I vary the normal routine by not having my bread with jam

DAILY OBSERVER

Dr Owl says... Show kindness, justice, equality and mercy to everyone

Tuesday, July 18, 2006 Banjul, The Gambia

D6M agric marketing centre for Banjulinding

by Ebrima Jaw Manneh

Dr Patrick Chang, Taiwanese Ambassador to The Gambia, yesterday inspected works at the D6 million funded Agricultural Produce Marketing Centre in Banjulinding, Kombo North. The one and half year-long project was jointly-funded under the Taiwanese Technical Mission (TTM) and the Action Aid- The Gambia.

Speaking to the *Daily Observer*, shortly after his tour of the giant marketing centre, Dr Patrick Chang expressed satisfaction with the work-in-progress, and then said: "It would be finally completed very soon. Most of the facilities have been completed and we will put in more equipment. The performance is good."

Dr Chang said the centre will serve as a powerhouse for the marketing of agricultural produce, including "vegetables, livestock,

Ambassador Chang

and rice. It can also be used as an exhibition centre for agricultural products".

According to him, the project would augment the income base of farmers and motivate them to stick to agricultural activities, saying it is a boost to agricultural productivity. "So, the project is for the whole country. This is the first one. We have started with Banjulinding, and hopefully it will be in the whole country," he said.

He said the TTM provided US$40,000, while Action chipped in US$120,000 for the project. "We provided an additional manpower, valuing D2 million. So, the amount for the project is D6 million," he said, adding that the centre has four refrigeration rooms that can keep the products afresh.

Tawanese Ambassador patrick Chang pointed out that the newly constructed Agricultural Produce Marketing Centre will be provided with management committee to oversee its day-to-day operations.

Dr Chang ended his inspection at the Banjulinding Horticultural Garden, where farmers, assisted by members of the TTM, expressed their preparedness for the raining season.

President Jammeh donates D20,000 to St Therese

President Yahya Jammeh has donated D20,000 to the members of staff and students of St Therese Upper Basic School in Kanifing.

Confirming this to the *Daily Observer* on Saturday, Ms Ozono Jammeh, Principal of St Therese Upper Basic School said the donated sum was in response to President Jammeh's concern for the 'Education for All' initiative.

She said the presidential gesture has enabled them to organise a speech and prize-giving ceremony in order to award outstanding students of the school.

According to her, President Jammeh is sponsoring all the female students in her school, through the President's 'Empowerment for Girls Education Programme'.

Ms Jammeh then thanked President Jammeh for the gesture and pray for his "long life, peace and prosperity".

6 British zoologists end research on dragon

by Alhagie Jobe

A team of six British zoologists, including a computer expert, photographer and an engineer, have ended a two-week long research on dragon (locally called 'Ninkinanka') in The Gambia.

Speaking to the *Daily Observer*, Mr Chris Moiser, a Zoologist at the Parfell Animal Land (Zoo) Caracals, said they have also conducted research on a lizard, locally known as Armitage's King, which was founded in 1922, by the then Governor of The Gambia, Governor Ch-Armitage (1920-1927).

According to Mr Moiser, the research had given them the opportunity to meet several people, especially in Kiang West, where "we met people whose relatives had once seen a dragon". He described

Members of the research group

various stories about the dragon as interesting, and then narrated: "At Kiang West, we were told that a 'Ninkinanka' was seen about 10 to 15 years ago, with a range of 15 meters long. "We talked to many people in the area who had grand parents, some of whom were fortunate to see 'Ninkinanka', but they said they are afraid of it. It is still a belief that if a Black man sees a 'Ninkinanka', he or she will die but cannot kill a White man", he said.

Chris Moiser opined that their research gathered facts, confirming the existence of 'Ninkinanka' in The Gambia. "The animal is in The Gambia but we are yet to see one. We will do all our best to ensure that we get the full story of the 'Ninkinanka'. "Many people assured us that in areas, such as Foto Jallon Island, 'Ninkinanka' could be found there. We will ensure that we are well-equipped next time we come around," he noted.

Other team members include Richard Freeman, a zoologist, Lisa Dasley, a photographer, Dr Chris Clark, an engineer, Seizi Marsh, a computer expert, and Olly Lewis, an ecologist.

Hunt for Gambia's mythical dragon

A team of UK dragon-hunters is on an expedition in The Gambia to track down a mysterious creature known locally as the "Ninki-nanka".

A child's depiction of the ninki-nanka on the expedition's blog

Believed to live in swamps, the ninki-nanka appears in the folklore of many parts of West Africa.

It is described as having a horse-like face, a long body with mirror-like scales and a crest of skin on its head.

Team leader Richard Freeman told the BBC, evidence so far was sketchy as most people died soon after seeing it.

Mr Freeman, a cryptozoologist from the UK-based Centre for Fortean Zoology, admitted that the ninki-nanka's existence was "very far-fetched indeed".

Second-hand accounts varied wildly from it looking like a crocodile or a snake to having wings and spitting fire, he said.

> There seems to be this thing when you see the ninki-nanka you will die usually within a few weeks
>
> Richard Freeman

But he disputed a suggestion that the hunt was a waste of time and money.

"We didn't know any of this before we came. We have to look into everything to see if there is a possibility that there's a real creature there," he told the BBC's Focus on Africa programme.

Cryptozoology is the search for animals whose existence is disputed or unsubstantiated, such as the Loch Ness monster.

Herbal potion

The team have interviewed one eyewitness so far - a park ranger from the Kiang West National Park who lived to tell the tale of his encounter three years ago.

He described an immense animal 50 metres long by one metre wide that he watched for more than an hour before being taken ill.

He put down his survival down to a herbal potion given to him by an Islamic holy man, Mr Freeman said.

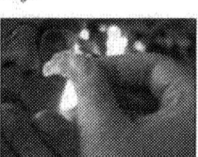
One of the expedition team examines a 'scale'

Later, according to the expedition's blog, after being shown pictures of various reptiles and mythical animals, the ranger said the creature's face most resembled that of a Chinese dragon.

"We've heard very similar stories all over The Gambia but mostly not first hand eyewitnesses... there seems to be this thing when you see the ninki-nanka you will die usually within a few weeks," Mr Freeman said.

The team are taking back a sample of what is claimed to be a ninki-nanka's scale to be tested in the UK.

But initially investigations suggest this is a red herring, perhaps a bit of rotten celluloid film and "not biological".

"We haven't discounted the possibility that there is a flesh and blood ninki-nanka in the swamps of West Africa, it's just at the moment the evidence is pointing to something more folkloric," he said.

DON'T MISS

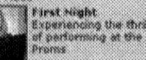
First Night
Experiencing the thrill of performing at the Proms

Escalating violence
The groups vying to become Israel's greatest enemy

Serie A sale
Match-fixing scandal could spark star sale
FROM BBC SPORT >>

on it, but having a banana in it. My northern upbringing comes to the fore – banana sandwiches for breakfast, but I enjoy them. As I finish, Richard and Lisa appear. I stay and talk to them for a few minutes and then realise that I should try and check whether or not the story about us is in the Observer. I make my excuses and leave Lisa and Richard.

The girl on reception can only suggest the Observer office as a place where I will definitely get a paper at this time in the morning. I walk up there, ignoring the bumsters shouting from the shelter at the other side of the road. They can shout "Excuse me" in a way that brings the average Brit to an instant halt; it takes willpower to ignore it. The Observer office is about 10 minutes walk away and I enjoy the walk, it is too early for the average bumster so I am not interrupted. The Indian pharmacist is opening his establishment and he waves with a smile - nice man.

At the Observer we have the security man sat in the chair outside, and another chap is with him with a bundle of newspapers under his arm. The security man has recognised me and he waves as I approach. We talk and I purchase a paper from the vendor, 10 D. is the price. A quick look reveals the story about us on the back page, together with the picture of four of us on the Observer balcony that was taken yesterday. I purchase another five copies, one for each member of the group. Due to the inevitable change shortage I ended up tipping the vendor.

The story was entitled "Six British Zoologists End Research on Dragon" and described what we had been doing in some detail, but a modest number of errors had crept in. It did, though, acknowledge our presence and work.

When I got back to the hotel there was nobody from the group in breakfast so I went and put copies outside each of the rooms. Ginger was outside Suzi's room, and promptly sat on her copy. As Ginger is a cat he probably saw it as his job. Some things do not change wherever you are in the world.

After showering, I spent most of the morning in my room, lying on the bed and reading. It was a matter of staying cool and conserving energy for what would be a long period of travel. I got up-to-date with the diary, and read the paper from cover to cover. At about 12.30 Suzi came and knocked on the door. The group were worried that I would fall asleep and be left behind. So I moved the bags, and myself, to reception where the *Gambia Experience* representative asked us to fill out a quality questionnaire. I asked about the aircraft departure from the UK; it was 25 minutes late, but they were still going to take us to the airport on time. Fine, 25 minutes is neither here nor there, it is when the delays run into hours, that the airport becomes less tolerable than the hotel. Our coach came on time at 1.00pm and we all trouped up to the entrance, with the gardening staff and a couple of receptionists carrying our bags. I gave my last £1.00 coin to one of the gardeners whom I had talked to during the week. The bus driver and a bus company baggage handler were also demanding a tip. As the baggage handler just lifted the case off the floor and put it in the coach, this seemed a bit excessive. I declined. I had to, I had no more currency. The trip to the airport seemed shorter than the trip in from the airport when we arrived. That is human nature, though, as the trip home is always shorter than the trip out. We checked in at the airport in separate queues. I think Dr. Chris got in first. I had worn one of my blue pilot shirts, and magically got an exit seat with extra leg room and an empty seat next to me – Oh bliss.

We waited in the departure lounge after a fairly routine check in and security check. The two Irish girls who had been staying in the hotel, came and talked to Lisa. One had an aunt who was a nun in Banjul. I misheard and thought that *she* was a nun. This was interesting, as she would have been the first nun that I had seen in a bikini. I don't have any odd interest in nuns, it just would have been different, and things always *are* different in Africa.

We boarded a bit later than the advertised time, with a crew who had been in the Gambia for a few days, and were airborne at 5.08pm UK time. The flight was remarkably uneventful, the films mundane, and the food boring. The landing was straightforward, straight in on approach, no circuits with other traffic. Our aircraft parked on an outer bay, and we were coached into the North Terminal where we went through immigration, and then waited for the bags before a totally uneventful walk through the customs hall. Suddenly it was over. There was a taxi driver waiting for Dr. Chris, Richard, and Olly. Lisa's son and husband were there, and Suzi had a courier who was going to drive her back to Devon. I said my good-byes, and made a couple of telephone calls before trying to settle down for the night. Jon was up in Woolsery. Richard had made contact already, so he knew that we had landed safely. I discussed a couple of points with him, and he let me know something about the media interest, and the fact that sponsorship for a future expedition was already being discussed with two possible sponsors. I don't immediately commit myself.

Wednesday 19th July

After something like two hours sleep, in twenty minute sessions, I eventually was able to get the early morning train out of Gatwick. It was 12.30 before I got back into Plymouth, after the obligatory change in Reading. A couple of hours sleep and a telephone call to Jon. He wants the whole expedition written up within a week to get this book out as soon as possible……..

Arial view of west Africa

Chris outside *Destiny's* nightclub

Richard on Bungalow Beach

The nightclub built next to Gambo's burial site.

Baka Samba, Ninki-Nanka killed his uncle

The team at Abuko

Crocodile pool at Abuko

The old pumping station, site of a sighting

Red colobus monkey *(Piliocolobus Sp.)* at Abuko

Vultures (*Necrosyrtes monachus*) in the hyena pound

"Ginge": The hotel Ginger Cat

Lisa on the shore

Verandah outside the hotel

Nile monitor *(Varanus niloticus)*

Hammerkop *(Scopus umbretta)*

Hassan Jinda with the old man who found the 'scales'

An acacia tree in bloom

Lantana flowers

A Fire Finch (*Lagonosticta senegala*)

Max, the bar owner

Giant bamboo

A large Baobab near Palma Rima

Chris Moiser at he sacred crocodile pool

Landing in Senegal

Mangroves on the Alahein

An African civet *(Civetticus civetta)*, the Gambian Reptile Park

A common baboon, the Gambian Reptile Park

Lisa holds an African egg eating snake *(Dasypeltis sp.)* at the Gambian Reptile Park

Luc Paziand: head of the Gambian Reptile Park

Toasting the memory of John Downes (sr.)

The rain soaked streets of Mandinari

Local children at Mandinari

Riverside near Bungalow Beach

The habitat of Armatige's skink

Lisa Dowley

Lisa Dowley was born in Solihull in 1969. Shortly after this her family re-located to North Staffordshire where her parents were successful licensees for most of her childhood. These entrepreneurial skills seemed to have been passed on as Lisa has had a number of successful businesses including owning and running a 'Live Music Venue' for a number of years in Manchester. While at this venue she became involved with 'Stockport Ghost Society' and organized numerous events such as overnight stays at various haunted locations throughout the country.

Her interest in Forteana (and in her words), "the not too readily and easily explained", began at an early age - six, to be precise. This interest in the unusual was facilitated by programmes such as, *Arthur C. Clarks Mysterious World*, and *David Attenborough's 'Fabulous Animals'*.

She is currently studying for a certificate in archaeology, at Keele University. Lisa has travelled, lived and worked abroad in such places as Canada, Finland and Germany. In 2004 she became the Staffordshire Representative for the CFZ, and has been afforded the opportunity to write a number of articles which she considers a great honour and achievement as she was diagnosed with dyslexia/dyscalculus while studying at the age of 33. Her interests and hobbies include: all aspects of Forteana, traveling, risk taking, Gothic music, art and dress, collecting squirrels tears, photography, hiking, camping, and meeting interesting people and their perceptions

She still resides in rural North Staffordshire with her family and 7 year old son.

Eighteen months ago, while surfing the internet, I came across a rather interesting and unusual website – that of the CFZ. I was extremely taken by the site and what they had to offer by way of curious and unique subject matter, which was presented in a very down-to-earth, un-stuffy, readable format. It was just the sort of stuff that appeals to an ageing Goth such like myself with limited free time, but with a big interest in all things cryptozoologically random, esoteric, and, er - well - things in general that are not too easily explained really.

I began corresponding with the organisation, and was offered the position of Staffordshire Representative. This role afforded me the opportunity to help out when I could with events, and to write a number of articles for them, which they deemed suitable enough to be added to their forthcoming yearbook. For someone such as myself with no previous experience of writing at this level, and dyslexia to boot, I found this to be highly confidence-boosting, inspiring, not to mention an immense honour.

Then shortly before Christmas of 2005, unexpectedly I received a 'phone call from Richard Freeman, the zoological director of the CFZ, asking if I would be interested in being a member of the team on the next CFZ expedition. I was totally gobsmacked and honoured, and naturally I answered `yes`, without even thinking of asking where we were going. The intervening days turned to months, and time soon passed away, and before I knew where I was, I was on my way to Gambia, West Africa, the 'Dark Continent'; my first taste of a genuine expedition abroad with the CFZ.

What follows is an account of that expedition, intermingled with personal thoughts and views that were amassed in the two weeks that were spent in West Africa, on the trail of the Ninki Nanka, a mysterious dragon/serpentine-like creature, that has reportedly been seen in and around the mangroves and shoreline of Gambia. This was the headliner of the expedition, but there was more to the expedition than this (there always is, with a CFZ project I have come to learn). Other lines of enquiry whilst out there, were to ascertain, if possible, the identity of a creature that was washed up on Bungalow Beach some twenty three years ago, and to establish whether or not a small reptile know as the Armatige's skink, which is endemic to the Gambia, is still alive and thriving.

July 3rd

As I live some two hundred miles away from the Exeter office of the CFZ, my journey to West Africa started a day sooner than everyone else's, with a very tedious National Express journey from my home town in North Staffordshire down to Exeter City Centre. This took five and a half hours, with the only highlight being a slight uproar from some of the passengers on arriving in Bristol to closed public toilets!

I finally arrived at my destination and then proceeded to wait a further 15 minutes for Richard to meet me. However, knowing how Richard is, I wandered around the bus terminal to find him sitting at the wrong stop! We greeted each other, and made our way to the taxi rank. We just about made the short journey to Holne Court in one piece, as the cab we were in was, let's just say, less than road worthy; so much so that the driver could not get out and help with the luggage as he had no brakes! This he inadvertently displayed by almost reversing over Richard on Exwick Hill by rolling backwards. We dropped my luggage off and then Richard, Oll, and I, quickly made our way to the train station to pick up Suzi who was also to join the team. That evening we had a meal at a local night-spot called 'Time Piece', and made exciting conversation, as we discussed our potential plans of action and what to expect in Africa.

July 4th

That evening, the night train journey to Heathrow was uneventful, with just one change at Reading. We arrived at Heathrow shortly before 6.00 am, and then preceded to find Chris Moiser, another member of the team who had elected to travel to Heathrow the night before, and spend the evening there on a bench. The last member of the team to arrive was Dr. Chris Clark, who appeared with not a moment to spare, at the check-in desk.

At last the team was complete:

1. **Richard Freeman**, Zoological Director of the CFZ, not to mention bad joke expert.

2. **Chris Moiser,** Britain's Best Mystery Big Cat Expert, and knowledgeable on all things African, after travelling there regularly for the last twenty years.

3. **Dr. Chris Clark**, who describes himself as an engineer but in reality, is an Astrophysicist by trade with diplomatic immunity. Dr. Chris is a thoroughly entertaining and knowledgeable chap, with a mysterious and intriguing interest in Afghanistan.

4. **Oliver Lewis,** a degree-educated Ecologist, who is a permanent resident at the North Devon CFZ HQ.

5. **Suzi Marsh** who's many skills include being a computer expert and the newest CFZ member, but more importantly one of those rare breed of humans, a sincerely genuinely nice person.

6. This just leaves me, **Lisa Dowley,** archaeology student, first aid officer, and security consultant, amongst other titles (and most of the time the street-wise voice of reason).

We off-loaded our luggage, and then in the blink of an eye, we were boarding the plane for the six-hour journey to West Africa. I can remember the staff going through the safety drill, and of being handed a pair of earphones. The next thing I remember is being awoken by a loud prehistoric roar of a T-Rex, which belonged to the in-flight movie, the latest *King Kong* film. `How apt`, I thought, 'a film containing thirty foot reptiles'. After regaining my senses, not to mention my hearing, I realised that the time had literally flown by, and before long we were banking up around the coast of West Africa to begin our decent to Banjul Airport. Out of the window, all I could see were vivid hues of lush-looking mangrove greens, which where intersected by swirling and twisting serpentine-like blue curves. These were the mangroves and waterways of the Gambia, the places we had come to explore.

We landed at Banjul Airport, at what seemed to be its only landing strip, shortly after 2.15 pm. I lined up to disembark from the plane, and the first thing that struck me was the heat as I emerged from the plane. I can only liken it to the heat from an oven while checking on Sunday dinner - a very instant, intense blast! As I was ushered by the airport staff to the bus, which ferried us to the newly constructed terminal building, a huge citrus swallow-tail butterfly flew past my head, turned as if greeting me, and then proceeded to land a short distance away. The sights and sounds all around me were very different; the cicadas singing in the bushes, the vultures flying high over head, and the crowds of people waiting on the other side of the very basic passport control, their eager smiles masking their poverty. Everything was very different, yet exotic and exciting, all at the same time.

After the mad scramble that was `Customs & Immigration`, we were led to a coach, given hand fans and a bottle of water, and promptly driven to the 'African Village Hotel', which was to be our base for the next two weeks. We had arrived in Gambia at the tail-end of the A.U [African Union] Summit [1], this being evident from the various banners greeting various delegates from the many divisions which make up Africa, that were strewn along fences.

These banners were the backdrop to huge acacia trees; their flame-red flowers were striking against the newly half-built, and then seemingly abandoned buildings that, no doubt, were to serve the purpose of impressing the visiting delegates and, in all probability, strategically designed and placed to give a false sense of sound economy. Giant mango trees where everywhere you looked, their branches weighed down with large ripe mangos, seemly asking to be relieved of their heavy burden. Intermittently, all along the roadside were vultures ominously sitting and waiting, but for what? I don't know.

We arrived at the African Village Hotel, to the tune of locals shouting, *'Hello, hello lady, you here long?'* and locals asking for monkey for no particular reason.

The hotel, in its appearance, had long since lost the sparkle of its two-star rating. It

---

1. **The African Union** (abbreviated **AU**) is a international organisation consisting of 53 African member states. Founded in July 2002 in South Africa, the AU was formed as a successor to the amalgamated African Economic Community (AEC) and the Organisation of African Unity (OAU). Eventually, the AU aims to have a single currency and a single integrated defence force, as well as other institutions of state, including a cabinet for the AU Head of State. The purpose of the union is to help secure Africa's democracy, human rights, and a sustainable economy, especially by bringing an end to intra-African conflict and creating an effective common market.

was more of a residence, which gave the vibe of being totally worn out, and fed-up and, in my view, had - in a very distant past - seen far better times. We were shown to our rooms, which were very simple affairs, and when the porter switched on the light in my room it made no difference whatsoever. However, in the dim, dark diginess, I could make out that my chalet had two rooms; I gave it a quick glance over, placing the hole-ridden mosquito net in the rickety wall cupboard, and thought to myself, 'well at least I have a bed'. I had nothing to compare it to, so my overall opinion was, that although it was very basic, it would do.

After being shown to our rooms, we all wandered down to the beach. How pretty it was too, with the warm Atlantic waves lapping over the black sands, in the heat and brightness of the day. While at the beach area we saw a variety of the indigenous creature of Gambia; cute little lizards called agamas, five-lined skinks, and countless beautifully marked butterflies. We spent our first night very informally sitting outside our rooms on the veranda, and it was not long before our conversation turned towards Bungalow Beach, the last known resting place of `Gambo`. We expressed our thoughts as to what exactly Owen Burnham may have seen washed up on that beach 23 years ago.

I went to bed on that first night wondering how much the shoreline may have changed, and of how best a group of six white English people could dig a hole without attracting too much attention.

July 5th

Breakfast was a simple buffet-style affair, with beans, fried eggs, French bread, mango, pawpaw, cornflakes, and a choice of tea or coffee. We took our fill, and after this, we made our way on foot down to Bungalow Beach. We walked along the main road, and then turned off down a lane that brought us on to the beach, where we walked along the shoreline to see what the tide may have left behind; there wasn't much really, a few sand dollars [1], a puffer fish [2], and a large amount of cuttlefish. After a short while we were joined by one of the many bumsters that inhabit the beach area, but I elected to ignore the individual and not get into a conversation with him.

---

1. Sand dollars are in the Echinoid (Echinoderms) class of marine animals. When alive, they are covered with a suit of moveable spines that encompass the entire shell. Like its close relative the sea urchin, the sand dollar has five sets of pores arranged in a petal pattern. The pores are used to move sea water into its internal water-vascular system, which allows for movement. Sand dollars live beyond mean low water on top of or just beneath the surface of sandy or muddy areas. The spines on the somewhat flattened underside of the animal allow it to burrow or to slowly creep through the sand. Fine, hair-like cilia cover the tiny spines. These cilia, in combination with a mucous coating, move food to the mouth opening which is in the center of the star shaped grooves on the underside of the animal. The anus is also located on the bottom, near the edge. Its food consists of plankton and organic particles that end up in the sandy bottom.

2. The *'pufferfish'*, also called blowfish, swellfish, globefish, balloonfish are fish making up the family Tetraodontidae, within the order Tetraodontiformes. They are named for their ability to inflate themselves to several times their normal size by swallowing water or air when threatened; the same adaptation is found in the closely related porcupinefish, which have large conspicuous spines (unlike the small, almost sandpaper-like spines of pufferfish). The scientific name, *Tetraodon*, refers to the fact that they have four large teeth, fused into an upper and lower plate, which are used for crushing the shells of crustaceans and mollusks, their natural prey.

From various conversations, I gathered that there has been a great deal of development along this stretch of beach, such as hotels and, not surprisingly, instead of 'X' marking the last resting place of `Gambo`, a huge nightclub by the name of *Destiny's* sat right over the site, or to be more accurate, the beer garden of the aforementioned club. This club is owned by the brother of the President of Gambia, who, as I write this, is currently under investigation for various crimes, most notably, fraud. In order not to attract too much attention, we acted as tourists, which served us well in being able to take photos of the area. However, there was a tense moment, when one of the staff (we assume) made some enquiries as to what we were doing. It became apparent that there would be little or no chance of asking anybody *if* we could dig a hole next to the club let alone dig up the beer terrace, so we took our pictures and moved on in a bid to explore more of the beach area.

We came upon an area called Kotu (meaning creek or stream) Market, (where a number of years ago Chris Moiser had the mind to do some er... how shall we say... networking. He had purchased some jewellery said to resemble Ninki Nanka from one of the stalls, the owner of which was a bloke named Baka Samba. He remembered Chris M, and we were able to secure a good deal on some trinkets. While there, Richard was making enquiries regarding Ninki Nanka merchandise, but all they had were stylised Chinese dragon carvings.

The conversation soon turned to the mysterious creature washed up all those years ago, and Baka (known locally as Mr. Fixit), who turned out to be very knowledgeable in our subject matter, was adamant in recalling `Gambo` as a dolphin, but with its dorsal fin absent. This man was a hive of information and said that when the dolphin was washed up on the beach it was still alive and that a number of white men had tried to save it, but to no avail. The dolphin, he said, then vomited and died and was then promptly buried on the beach.

He also relayed to us a story. Many years ago, as a hunter, his uncle had actually seen Ninki Nanka. However, his description was not very detailed and he only went as far as saying it was very, very big, scary and terrible! He did go as far as to say it had four legs, a long tail and a horrible head which had fire coming out of its mouth. He was also very adamant that this creature was very dangerous and that most, if not all, who saw it would die within a time range of around four to five years. At this point Baka's nephew joined in the conversation saying that Ninki Nanka resembled a large crocodile but that its eyes, teeth and head were different. When asked if he had seen the creature he nervously laughed and said *"No, I do not wish to see such things!"* We did manage to capture most of this interview on video.

We took our newly purchased goods, thanked all for their information and made our way to another hotel for some well-needed refreshment. While there Richard was approached by a man who said he knew where to find the Ninki Nanka. According to him, the location appeared to be some great distance away, but, although he could not give a price on the cost of such a journey, we took his number anyway. Some of us then got a taxi back to the hotel as the heat was proving too much for some, but Rich, Dr. Chris and I walked back where we all met up in the pool and discussed the day's information. A couple of us then assembled the day's blog to be sent back to England. In the bar later that evening we met a guy by the name of Ian Clifford who is a medical photographer by trade, but was in Gambia hoping to photograph mambas.

Later that night, I looked over my purchases of two Ninki Nanka pendants and on closer inspection I could not help thinking that they were nothing more than stylised Chinese dragons.

July 6th

Today we decided to make our first visit to Abuko National Park; this area being the setting for one of the most well-known tales of the Ninki Nanka. We left the hotel shortly after 9.00 am when, at the end of the long driveway, awaited Chris M's long term associate, Assan, who ran a small taxi company. Assan was to become our trusty driver and protector from the con-artists.

The journey to Abuko National Park took just under an hour and, although a dusty and bumpy commute, was full of interest. We had a real glimpse of the surrounding area en-route; the corrugated shop fronts that also doubled up as homes, the street corner vendors plying their wares of nuts and fruit, and the women in their bright and colourful traditional dress.

We were dropped off at the gates of the park and agreed with Assan a time for him to pick us up. Chris M paid our entrance fee, which amounted to the exorbitant sum of 63p.

The second we entered the park our eyes were drawn immediately to the abundant wildlife; millipedes were all over the dry dusty path and set back in the jungle foliage we could make out huge termite mounds. All manner of vines and creeping foliage filled my line of vision - it truly was a different world.

We made our way to the first hide (Darwin Field Station), which was a huge hut mounted about 12 feet off the ground and looking over one of the many man-made pools. As we were drawing near the end of the dry season, the water level was very low and these pools are inhabited by, amongst other things, Nile crocodiles. After some wait we saw them, two in fact, at either end of the pool, much to the joy of Richard, as this I think is his favourite animal. We saw a number of other creatures there also: a malachite kingfisher, a monitor lizard, and a couple of red colobus monkeys playing in the tree canopy above the pool. As we were departing the hide we were joined by a guide. We were somewhat apprehensive about this as we were beginning to tire of so-called guides just out for our money, but we had no need for concern as Musa Jatta, as he was known, turned out to be an exceptional guide. At one stage his bird calls were so convincing I had trouble working out which were his calls, and his response from the birds was amazing.

With Musa taking the lead, we began to follow the trail, and he proved himself to be an excellent spotter - in fact at one point on the trail he leapt back outstretching his arms and shouted *"SNAKE!"* Rich stopped abruptly causing a sort of domino effect with the group, and as I manoeuvred round Richard in order to see, I heard him gasp *"Wow!"* What I saw was an amazing sight, for there in its natural habitat was a juvenile forest cobra no more than two or so feet long. It must have been disturbed by our approach and had taken exception to this by spitting at us; it turned away as if to slither off and we took half a step forward and it must have thought better of it and turned and spat at us once more, before disappearing into the jungle undergrowth. I found the whole 10 seconds breathtakingly mesmerising. The fact that we had seen jacanas, pied kingfishers, cordon bleus and ground squirrels just did not compare to the adrenaline buzz of seeing a spitting forest cobra.

To make conversation, Rich spoke to Musa about the Ninki Nanka, and *his* interpretation of the creature was a myriad of different animals. To him, it was a huge python with legs and webbed wings similar to that of a bat. It could transverse over land and water, but he did not know if could actually fly. He also went on to tell us that, in the early part of the century, during heavy rainfall, a dragon had caused a lorry to spin out of control, and crash as it was crossing the road.

Meanwhile Chris M had gone ahead of us to the café at the half-way point, which I noted was aptly named the Ninki Nanka Café, where he had struck up a conversation with a woman stall-holder there. He gained more information about the famous sighting that occurred there in the 1940's and also that the grandson of Papa Jinda, the man who had seen the Ninki Nanka, all those years ago, was working at Abuko and went by the name of Hassan.

While making our way back to the entrance we passed through the remains of the old pumping station. This is where Papa Jinda had been working when he saw the Ninki Nanka. The whole area was just totally surreal to me, being a mix of the Palace of Angkor, complete with strangulating vines and trees crushing and consuming what was left of the buildings, with a dash of a haunting purposively-constructed Hollywood film set about it.

It truly had a beautifully eerie and mysterious sense about it.

As we made our way back to the entrance, a brightly patterned juvenile monitor lizard resting on some fallen palm leaves bid us farewell.

It was at the entrance where we found Hassan Jinda, a tall man in his mid-to-late twenties. He relayed the tale of his grandfather's encounter. Hassan was adamant that his granddad had seen the creature on two occasions, once in 1943 and then again in 1947. Mr. Jinda Sr. was at the pumping station (when it had been working then to supply water to the surrounding area) in Abuko, and witnessed Ninki Nanka cause damage to a number of pipes, thus disrupting the water supply. The local workers had become fearful, and had demanded that a mirror be erected in order to scare off the beast, as it was believed that the only sure-fire way to rid the area of the creature was for it to see its own reflection.

At the time, Mr. Jinda Sr. described the creature as being huge with simmering diamond-like scales, and a fire-crested laden head. He believed the creature he saw to be female as it was locally believed that if you saw a male you would die instantly. The second time he saw the creature was in 1947, but after he had come across it this time he became ill and suffered pains in his legs and side. His hair suddenly began to fall out, and within two weeks he had died (this is one of the few consistent themes to the folklore relating to the Ninki Nanka death within a set period of time).

After Hassan had told these two stories he then went on to say that he knew a man who possessed what he thought were Ninki Nanka scales. I'm not totally sure what transpired next as its seems a bit of a blur, but as clichéd as it might sound, before I knew what I was doing, I was making my way back into the park with Richard and Hassan in a bid to find the man with the said scales. Some two hours later we re-emerged at the entrance only to find the scale owner sitting there waiting for us! `Bloody typical` I thought.

As is the custom, and in order to be polite, we ordered some drinks for everyone, and the team all gathered round and sat down, as the old man produced a pouch from around his neck. As he opened it and I heard the contents hit the cardboard tray, I instinctively knew that they were not of biological/organic origin. In fact, they sounded almost glass-like as the pieces chinked together upon hitting the surface of the tray. In colour, I could only describe them as a dirty silvery, almost clear, opaque.

We all looked at a large sample, and collectively we hid our disappointment well so as to not offend anybody; to this weathered old man these were Ninki Nanka scales, and he used them to make charms. As it seemed only proper to obtain a sample of what he was purporting to be a Ninki Nanka scale, in order to test it once we got back to England, we asked if we could buy a piece. An agreement reached, at the princely some of 500 Dalasis, I placed the article into a sample bag. We then bid our farewells, and thanked everybody for their time, before making our way back to the hotel.

Over dinner it was discussed and decided that, on Saturday, the team would make the three hour plus journey to Kiang West in order to follow up some leads regarding the Ninki Nanka.

Whilst in my room that night, I pondered over the information we had amassed, the seemingly varied and confusing descriptions of the Ninki Nanka, and the display of limited education which is, for the most part, overshadowed by the cultural beliefs of the locals. Do they believe that what they had were indeed scales from a Ninki Nanka, and was their belief in such as creature as genuine as their fear for it? Or I wondered, did they just pull a fast one in pursuit of a quick Dalasi on some gullible looking 'whities'?!

July 7th

During the early hours of this morning (around 4.00am) I awoke, or rather my stomach woke me up, with some ominous lower bowel grumblings; yes, I think I have the making of the famous, or should I say infamous, 'Banjul Belly'.

I could not get back to sleep, so I went down to the veranda area to catch up on my notes and to watch the sun rise. To my surprise, I was not the only one who could not sleep, as another hotel guest - an elderly lady by the name of Julianne - was there chatting to the security guard, Sueliman.

You know, it's amazing how you can meet the most interesting of people at 4.30a.m. Julianne, who must have been in her late 60s or early 70s, turned out to be a very well-travelled individual. This was her fifth time to the Gambia this year, as she was sponsoring a young man here. Prior to this she had journeyed up the Amazon, and met indigenous tribespeople there, and had been introduced to their sacred anaconda; this was just one of the many tales of her many other journeys throughout the world. Wow, I must admit to being somewhat envious.

The conversation got around to what I was doing here, and I duly explained. Sueliman began to explain his understanding of the Ninki Nanka, and he relayed in his description that the creature had some form of crest, (which had been mentioned in other accounts) but more interestingly he added that under, on, or to the side of this crest, is some form of text (Islamic). *"If you see this writing you will die"*, he muttered in a very low, quiet, serious voice (he had never seen it, I hasten to add).

He also said that the scales on the upper front part of the creature, were 'rainbow-shiny' (his words not mine) comprising of a mainly red/orange colour.

I returned to my room (mostly because I was in dire need of the loo), and got ready for breakfast, although I did not eat much that morning. When Richard awoke, I relayed the information that Sueliman had told me, and I told him when he would be at work next, as he only did nights (I had arranged for Richard to interview him at a later date, during our conversation).

In light of the long and productive day we had spent at Abuko National Park yesterday, it was thought best that Friday should be spent as a more casual affair (some of us having trekked round the park twice), so that the group could rest up ready for the long trek that lay ahead on Saturday. I found this an excellent idea, as I was beginning feel the effects from what is medically known as 'Banjul Belly'; thankfully only a mild case, which I can only liken to watching a terrific storm out to sea, and praying that it does not blow inland.

Whilst at breakfast, it was discussed that we may venture down to the local fish-market, as it is not much more that a 10 minute walk away (it can be seen from the veranda of the hotel, being situated on the beach just down from us), to see if the local fishermen's catch of the day had anything unusual, or of interest to offer. However, firstly, due to the large amount of information that had been amassed at Abuko, we all set about making notes and comparing information.

While Oll was busy with his blog entry, and after we had finished documenting our information, Richard, Dr. Chris, and I, went down to the sea for a swim, and to have a look around the shoreline for anything of interest. This 'Atlantic' refreshment was greatly needed, as today there is no cloud, which is making for a very bright day with intense heat; I also found it soothing to a certain degree with my stomach complaint! Dr. Chris and Richard wandered off to a large outcrop of rocks in search of potential things of interest, while I had a look along the small shoreline. While doing this I came across a cuttlefish. At first glance I thought it may have still been alive, as its chromatophores [1] were still active, however they can carry on working for some time after death. I named him 'Cthulu' [2] Although dead, it was nice to see such a creature in its entirety, rather than stuck in the side of a budgie cage.

After returning from our Atlantic refreshment, Suzi, Oll, and Chris M, were off to the internet café to send off the latest report. Richard and I went to look out on the veranda to see if there were any signs of life at the fish-market; as we looked out we could see that the market place was bustling with life, so we decided that we would leave a note for the other guys for when they returned, letting them know of our whereabouts if they wished to follow us down. Dr. Chris, Richard and I gathered what we needed and made our way down to the fish-market, which was a straight 10 minute walk via the main road just outside the hotel.

Now, when I say straight walk, this implies that you do not get accosted by the locals

---

1. Chromatophores are pigment-containing and light-reflecting cells found in amphibians, fish, reptiles, crustaceans and cephalopods. They are largely responsible for generating skin and eye colour in cold-blooded animals and are generated in the neural crest during embryonic development. Mature chromatophores are grouped into subclasses based on their colour (more properly "hue") under white light: xanthophores (yellow), erythrophores (red), iridophores (reflective / iridescent), leucophores (white), melanophores (black/brown) and cyanophores (blue).

Some species can rapidly change colour through mechanisms that translocate pigment and reorient reflective plates within chromatophores. This process, often used as a type of camouflage, is called physiological colour change. Cephalopods such as octopus have complex chromatophore organs controlled by muscles to achieve this, while vertebrates such as chameleons generate a similar effect by cell signaling. Such signals can be hormones or neurotransmitters and may be initiated by changes in mood, temperature, stress or visible changes in local environment.

2. Cthulhu (other spellings: *Kutulu, Kthulhut, Thu Thu, Tulu,* and many others) is a fictional entity created by horror author H.P. Lovecraft. *Cthulhu* is often preceded by the title *Great, Dead,* or *Dread*. Lovecraft suggested that *Cthulhu* is pronounced "Khlul'-hloo" S. T. Joshi points out, however, that Lovecraft gave several differing pronunciations on different occasions. According to Lovecraft, however, this is merely the closest that the human vocal apparatus can come to reproducing the syllables of an alien language.

Cthulhu debuted in Lovecraft's short story "The Call of Cthulhu" (1928) - though he makes minor appearances in a few other of Lovecraft's works. August Derleth used the creature's name to describe the system of lore employed by Lovecraft and his literary successors, the Cthulhu Mythos.

The most detailed descriptions of Cthulhu in "The Call of Cthulhu" are based on statues of the creature. One, constructed by an artist after a series of baleful dreams, is said to have *"yielded simultaneous pictures of an octopus, a dragon, and a human caricature.... A pulpy, tentacled head surmounted a grotesque and scaly body with rudimentary wings."* Another, recovered by police from a raid on a murderous cult, *"represented a monster of vaguely anthropoid outline, but with an octopus-like head whose face was a mass of feelers, a scaly, rubbery-looking body, prodigious claws on hind and fore feet, and long, narrow wings behind."*

who hang around outside the hotels wanting to be your guide, which is what happened to us. We were besieged by many-a-local, so a 10 minute walk became somewhat longer. One of them persevered and clung to us. We did not obtain this guy's name, as we were trying not to encourage him; if you do not require a guide then the rule is to politely ignore them, and they will eventually drop back, disappear, and seek out their next tourist. However, Richard made the mistake of conversing with him, and we eventually arrived at the fish-market with an extra member in our group.

The market was full of people going about their business of cutting, drying, salting, and smoking, various kinds of fish. We made our way toward the pier where the larger types of fish were being kept in a manner that would have caused the UK's Health & Hygiene officials to have a breakdown, but it must be remembered that this is Gambia and things are done differently here; people make the most of, and utilise, what resources they have at their disposal.

The rusty, battered freezers were lined up all along both sides of the pier (not plugged in, naturally). Wherever there was a gap void of a freezer, families crouched on the floor, eating together. In some spaces, individuals slept after a long day's fishing, and mothers nursed and cleaned their children. Richard asked a local fisherman if he could look inside the freezers to see what they day's catch had brought home.

On the whole there was nothing of major interest contained within the freezers; mainly just a mixture of ice, sole fish, butterfish, and barracudas, but because Richard had sparked up another conversation, we now had a second person joining our group who promptly took it upon himself to give us a tour of the market. In effect, we now had stereo guides, (not an ideal situation I thought), who Richard unwittingly facilitated by asking more questions as he was so caught up in his quest for information.

At this point I began to start feeling a little uncomfortable with the developing situation, as this second 'unofficial' guide led us to the less busy, smokehouse area of the market. While *en-route* to this part of the market, I was approached by a man, who leant over my left shoulder, and said, in a low voice: *"My name Saul. I work African Village Hotel"*, at which point he showed me his pass-card. *"Be careful, these men dangerous"*. On hearing this, I felt my suspicions vindicated and discreetly told Richard what had been relayed to me while maintaining a smiling face so as not to alert our unwanted guides. I could not believe what he did next; to my 'horror' Richard said `OK`, and promptly carried on conversing with the two men. I was instantly filled with rage, firstly for being ignored, and secondly for the fact that Richard was totally oblivious to the potential dangers of the situation that he had gotten us into.

I now began to take on a different approach to the whole situation, and was not interested in any of the conversation that was going on. My main objective, as I saw it, was to get all three of us back to the hotel a.s.a.p, without too much hassle, and in one piece. I casually mentioned getting back, and we started to make our way out of the fish-market and up on to the main road. At this stage, I was at the back of Richard and Dr.Chris, observing what was going on around us, and looking for any signs that might have indicated a potential situation. However, while making our way back to the main road the fisherman had started to tell Richard of the need for cement to carry out repairs. Richard believed the man's story, as he found it refreshing that the guy was not asking for money, and thought his story credible, and had agreed to buy cement!

We finally reached the main road, and, as I thought, averted any potential danger; alas, I was wrong. We crossed over the road to head down a rather dingy and, not to say, dodgy looking alley. The only thing this alley was lacking was a soundtrack similar to those you get in B-rated horror movies, you know the sort, and the music which implies something nasty is just about to happen!

I asked Richard, *"Where the hell are we going now?"* He replied, *"To get some cement"*. I replied: *"I'm not going down there! I'm going back to the hotel"*. I was thinking that the guys would see sense, or get the fisherman and the other local to bring the cement to them. However, to my jaw-dropping horror, they followed them down the alley and disappeared to the left.

I began to make my way back to the hotel, thinking how on earth do I explain that I've lost the CFZ's Zoological Director and an Astrophysicist, when I became aware of someone shouting at me. However, thinking it was another local trying to make a quick dalasi I ignored it. The next thing I know, I am being tapped on the shoulder by a Gambian policeman who looked very concerned. The man that had first warned me, had apparently got one of his friends to ring the police, and the policeman asked in a very concerned manner, *"Where have your friends gone?"*

I explained to him that they had gone down the aforementioned alley, and that I had refused to do so. He replied "good lady", and went on to say that the man, meaning the fisherman, was very dangerous and bad, and that bad things may happen to my friends. *"We must find your friends immediately, come with me"*, he exclaimed in a very adamant, concerned manner.

Even with an official I was still rather hesitant to go venturing down the alleys and back streets of Bakau. We searched the corrugated shack-filled alleys, asking the locals which direction the two had been taken in for some time until we found them, at which point I hung back, while the officer went over and fetched Richard and Dr. Chris, who were still seemingly totally unaware of the potential danger they had put themselves in.

The officer explain loosely (as by now a small crowd had gathered) regarding the fisherman's *modus operandi* and light-heartedly said: *"There are good and bad people wherever you go, in future listen to the lady, she has sense"*. Needless to say, the moment the policeman appeared on the scene, the 'fisherman' disappeared.

I felt very motherly at that moment and I was glad they were OK, but by God I wanted so much to tell them off, and give them a lecture of the dangers of strangers.

Thinking the worst was over, we began to make our way back escorted by Babu, the man who had telephoned the police, but again to our dismay, he also began to try and obtain money, by mentioning that he had got married yesterday, and why hadn't we been at the wedding?

Wanting to get back, I took matters into my own hands. Being ill earlier that day, I explained that I was unwell with 'Banjul Belly', and exaggerated my condition, which, in light of the situation, I felt was justified, and needed to get back to the hotel to take my medication. Feigning this illness served us well; so much so that Richard thought that I actually *was* severely ill. We gave Babu a few Dalasi for his troubles, as after, all he had called the police, which stopped a potentially serious incident from occurring, and made our way back to the hotel, where we re-told the afternoon's events to the rest of the group.

Richard was very quiet for some time, then apologised, and said that he had been caught up in the moment. I think that the moral of this story for today children, is that sometimes adults should not talk to strangers either, and if they do, they should think long and carefully before doing so. All in all just another day on a CFZ expedition!

July 8th

Again I was awoken with my 'Banjul Belly' in the early hours, so decided to stay up and make sure that the guys had got all they needed for the day trip out to a remote village by the name of Dumbutu, Kiang West. I had made the choice of not going on this venture, due to my complaint, as I did not think it would be fair or fitting for I may hold and hinder the group on the day's excursion. What I had done in order to contribute, is help to out with the deposit, and loaned my camera to Suzi, and my dictaphone to Richard. Chris M did not go on this venture either, as he was also suffering somewhat. I stayed behind, and with the aid of Suzi's laptop, began to write up the previous day's events for the blog, but this took longer than expected as the power was very intermittent, and at one stage I switched the laptop off as a safety measure.

I spent most of the day in the murky depths of the hotel pool to ease my tummy, and chatting to Chris M. We agreed not to get concerned for the well-being of the team until after 8.30pm, at which point, while Chris and I were sitting in the TV area, the gang duly arrived, full of conversation of the day's events which they relayed over dinner that evening.

July 9th

The team spent today in a more relaxed manner. After their gruelling journey yesterday, plus they had amassed a great deal of information which had to be documented, while still fresh in their heads. We discussed what to do next, and I suggested that if we were going to Banjul, the capital, why don't we check out the museum to see if they had any historical information pertaining to Ninki Nanka? We also arranged to go down south of the country to check out a sacred crocodile pool at Cartong, and possibly check out the habitat where two Armitage's skinks had been found. En-route, we were going to check out the Reptile Park, and hopefully get out on to the Alaheen River, which separates Gambia from Senegal. I was looking forward to this as I was beginning to get a little stir-crazy, having being held up at the hotel for the last couple of days, as you really could not go for a walk outside the hotel complex, as the bumsters' pestering at times can be really annoying, and too much to bear at times.

July 10th

We arrived in Banjul mid-morning, and to my surprise it was almost relatively hassle free; we just came across the odd beggar who seemed genuine as, in a couple of cases, they were disabled in some way or another. The city was pretty much how I expected it to look, with run-down remnants of buildings left over from colonial times mixing in with new buildings, and with the odd shanty-style hut attached here and there. Everywhere, everybody was going about their daily business.

Whilst in the capital, as he was familiar with the area, Chris M took the lead. He was looking for the recording of a song he had heard some years ago about the Ninki Nanka, so we made our way to Albert Market, where Chris disappeared up a flight of steps. I suggested that the rest of us waited downstairs, and I gave Chris M a set period of time before we would become concerned, but, thankfully, before too long, he appeared from out of another doorway, rather pleased that he had in fact managed to get the song he was looking for.

After this, we re-traced some of our steps, and headed for the museum. On arriving we were told to leave our bags at the desk, and were not allowed to take any pictures, so we all took turns in standing by our bags. I genuinely felt sorry for this building; it was rather dusty and worn out, and - pretty much like Gambia - it had in the past seen far better times.

It had no information regarding the Ninki Nanka; instead the rooms were filled archaeological finds from the many pre-historic areas in Gambia such as the stone circle at Wasso.

There was also a great deal of information regarding colonialism. The only folklore to be found was in the form of a story from Ghana, when a king had made a pact with a giant snake that brought the rain. If it was given young girls as food, it would ensure that good rains would come for the crops. This snake lived in a well and the story transpired that one of the king's sons fell in love with one of the sacrificial victims and, therefore, killed the snake in order to be with the girl. We asked the staff at the museum if they had heard of the Ninki Nanka, but they hadn't. In fact, I thought the staff were most unhelpful, and very un-knowledgeable about most things they were asked. Instead, they ushered and pointed us in the direction of the souvenir shop, which came as no surprise as this was a government owned building, and as such I should imagine that their wages were low.

We left the tired, worn out, museum behind, and Chris M led us to the sumptuous Atlantic Hotel (it made our two star hotel look like an open sewer), where we had some much needed refreshment, and drank a toast to Mr. J.T. Downes Sr. In the grounds behind this hotel, is a small bird garden-cum-reserve, so we wandered through, but for some of the team, the mosquitoes proved too much and they returned. Rich and I, however, ventured on through, and were rewarded buy seeing a colony of epauletted fruit bats having their daytime sleep.

We returned to the hotel in the afternoon, and I went with Chris M to the local supermarket, to buy some local foodstuffs to take home. Later, we all sat in the pool discussing our various views on the day's events. At dinner we were joined by Ian who had been to Abuko in order to film snakes, but had not had much luck.

July 11th

Today was Olly's birthday. We had made plans to go south of the country today, but our trusty driver, Assan, could not make it today. So we spent the day not doing a great deal, which I thought was a bit of a waste, and, in the most part, very boring. I have never been on an expedition before, so I have nothing to compare it to but I should imagine that sometimes you get days like these, and the best you can do is go over your notes.

July 12th

After feeling like a caged bird yesterday, I was chomping at the bit to get out and explore more of Gambia. We hired two cabs, and made our way south to the Alahein River which separates Gambia from Senegal. On arrival, I purchased drinks all round, and brushed up on my French (as the guy who owned it was, yes you guessed it, French). While drinking our fruit cocktails, the local fishermen were busy trying to organise two dug-out canoes, so we could get out on the river

Being out on the River Alahein was one of the most serene relaxing things I have ever done, to the left of us was Senegal, and every so often I would outstretch my arm and touch the mangroves on that side. It was high-tide (which made the water below us around four foot deep), so we headed upriver. While on the river, you could hear the constant roar of the distant Atlantic Ocean. We then turned round an island that is popular in high season with tourists for its wildlife/birds there.

Whilst walking about, I saw juvenile mudskippers, hermit and fiddler crabs. Everywhere you looked there were piles of shells, and it was explained to us that these were used for whitewash. A particular pile caught my eye as it had a stick atop, with various beads and adornment; this was juju. It had been placed there to warn off other people from stealing the shells. We were ferried back to the main shore of

Gambia, and made our way to Folonko, a village in Kartong, to visit the sacred crocodile pool.

Whilst at this site, we were soon surrounded by the village children who were very pleasant, and did not ask for anything, but to see our cameras and have their picture taken. We had to remove our shoes before venturing near the poolside, as it was a holy site. A young man appeared, and took the role of our guide, explaining the role of these pools; when the women wanted to bear children they would come down to the pool's edge, to take water, and bathe in it, as it was thought to possess fertility magick. He also showed us a small hut that was used for the performing of female circumcision, which is widely practised throughout Africa. I found it strange that a religion such as Islam, which forbids the worship of any other idol, to be as tolerant of such practices as crocodile worship, but this practice is very ancient, and is part of the cultural heritage, and so forms the identity of these people.

The pool itself was frighteningly low on water, and I found it hard to believe that there were in excess of twenty crocodiles in there hiding under the water hyacinth. What I thought at first to be a tree root on the far side of the pool, turned out to be a young crocodile, and I quickly took pictures of it basking in the sunlight. We were also told that there was a white crocodile in there, and to that to see it was good luck. Richard got round to asking about Ninki Nanka, and the guide told us of their belief that it was a snake that grew into a huge python, which, as it entered the sea, in turn, changed into a dragon.

At this point, one of the teenage boys in the group told of a song that the men sing while they are bathing after male circumcision. It is sung to ward off the said creature, and he then promptly broke out into this song. We gave him a round of applause, before leaving and, quite willingly, gave them money for their time and effort, not to mention how well-behaved they all were.

Our next stop-off point was the Gambia Reptile Park, which is owned and run by a very charming and likeable Frenchman by the name of Luc Paziand. It is unlike any other park I am likely to visit; it consisted of concrete pits set into the ground with, in most cases, a 3 foot high wall above ground, making the total depth of most pits around 4 foot. You just peer over the edge of the wall, and into the pits' rather shallow depths, to be greeted by an assortment of venomous reptiles, with no barriers or glass partitions, but just your common sense as protection. I found the whole experience very refreshing.

We were given a tour by a very young girl, who was very informative on all of the creatures there (ball pythons *(Python regius)*, African egg-eating snakes, *(Dasypeltis scabra),* spitting cobras *(Naja nigricollis)*, puff adders *(Bitis avietans),* monitors, skinks, burrowing cobras *(Paranaja multifasciata)* and a few non-reptile inhabitants that had been rescued and were in rehab ready to be set free into their local habitat. These included a rather cute and loving pelican, and a not so cute looking baboon. At the end of the tour we asked if Luc was in and if he would talk to us. He was and we chatted for some time in the grounds of the park; more so when he discovered that

Richard was experienced with reptiles, and he then invited us to take pastis and water (similar to Pernod).

When we arrived at his home, we were greeted by his highly active and entertaining son, who must have been no older that two; his antics and acrobatics where frighteningly confident for somebody so young. Luc relayed that the other day, his son was found on top of his partly constructed roof, which is around 20 feet high, and last week he was dancing round the well wall; the drop inside the well being around 60 feet!. I don't know whether it was the antics of Luc's son, or the heat, but the majority of the group where rather quiet, so I headed the conversation with Luc.

While drinking, he explained his work with Bangor University regarding the puff adder, as there is no anti-venom if you are unfortunate enough to be bitten by one; the best you could hope for is amputation, and the worst an agonising death. He was also working on various aspects of the spitting cobra's venom. He also told us that only until last week, he had had two Armitage's skinks in his care, but one had died from some existing injuries, and he had decided to release the other as sourcing its food had become too difficult. He told us that the best place to look for Armitage's skinks is the grassy dunes along Gunjur Beach, and also said that the Mandinkas referred to it as 'bauko kono sa' which means snake in the sand. I think this makes reference to its ability to quickly disappear beneath the sand. He said that he had some images that Ian had taken, and he was quite happy for us to have a copy. The subject got round to Ninki Nanka, and he had been told the now familiar story by locals that it was a huge snake, and that when entering the sea it changed into a dragon (I was beginning to think at this point that the Ninki Nanka was possibly nothing more than a giant sterile eel). Luc's explanation for the crest was quite plausible; he thought it could have been ticks or leeches stuck on the head, and that these had been misinterpreted as a crest or crown.

On a more serious note, Luc expressed that he was thinking of moving his whole operation to Guinea, as trying to run a business in Gambia was proving very difficult. Luc also stated that Gambia is only a democracy on the surface, the reality being a very different picture, but he would say no more on this subject. We made arrangements to obtain the photos he has of the skink and we said our farewells.

The third and final leg of the day's excursion, took us back to Bungalow Beach to meet a man called Lamu, who had a story regarding Ninki Nanka. He claims that his grandfather could summon the creature, and said that his granddad would come to Gambia and perform this ceremony for us. He was very lax in giving us a description of the beast, and instead invited us to Guinea where all would be revealed! I was sceptical of the whole story, even more so when he produced a scale from the creature. It looked more like a piece of glass than anything else, and I got the impression that Lamu did not take any of it too seriously, and was merely attempting to see his family.

During our evening meal, Rich was summoned to the main reception of the hotel to receive a 'phone call. When he returned it turned out that it was BBC World Service

(Africa), who had read of our exploits in one of the national papers back home in England, and wanted to conduct an interview with him. This he arranged for some time tomorrow morning, and I began to wonder just how much of a stir we were creating back home. At dusk, we also planned to visit the beach in Gunjur that Luc had mentioned, in the hope of catching a glimpse of the elusive Armitage's skink.

July 13th

Richard conducted his interview, which he felt went rather well. Then, very late in the afternoon (around 4.00 pm), we prepared, and gathered our things to travel down to Gunjur Beach. After a journey of an hour or so, and three police checkpoints later, we arrived at a rather smelly fishing-port, where the day's work had already been done. There were not a great deal of people around, and I thought, *"Great, no bumsters to bother us here"*.

Oh, how wrong I was.

After waking along the beach, we came across the beach bar that Luc had mentioned. This was owned by a couple of Rastafarian guys, and it was them who had found the skinks and taken them to Luc. This bar was full of Gambian character, with wind chimes made from all sorts of flotsam and jetsam, and discarded items from the fish-market, such as shells, hung from every space available; all adding to the atmosphere of the place. We ordered a round of drinks in order to provoke conversation. One of the men confirmed Luc's version of events; it was, in fact, he that had found the skinks in his toilet no less! As he had heard of the work Luc was doing at the Reptile Park, he had passed them on to him, and described them as being around 10 inches long, as thick as his finger and sandy-brown in colour with a black stripe.

Richard asked him about Ninki Nanka, and we got the now familiar response; he said that it was an animal so powerful that those who saw it died. He had known people who had seen it, but they were 'no longer conscious'.

We walked along the upper point of the beach as there seemed to be remnants of a path there; this afforded us a view of beach and dunes, the area that Luc had implied was the skink's habitat.

We were minding our own 'skink-hunting business' when we were joined by a beach bumsters with his dog. The last thing we needed was another bumster tagging on, let alone with his dog as well.

He followed us everywhere, and Suzi asked him to leave us in peace to do our work, but he just ignored her. Richard got around to asking him what he wanted, to which he replied that he was there to help, but I am sure this would have involved a nominal fee of some description! Richard explained that we were scientists in search of a very rare lizard, and that his presence, along with his dog, would only serve to scare away any chances we had. After some hovering on the bumster's part, he and his dog eventually disappeared.

We searched the dunes looking in many a hole and under rocks. I dug down into the dunes to see what lay underneath, and after a short time digging, I came across very damp, compacted sand at about 6 inches below the surface. I mentioned this to Richard, as I thought that the speedy burrowing of the skink would only be to a certain depth, and then, maybe, it would move underneath, and across, the dry layer of sand with some of it used as cover above. Surely, I thought, if it dug any deeper, it would slow up any creature. We wandered quite a distance along the Gunjur Beach shoreline, and as the evening began to draw in, we sat in what we thought may be an ideal area for skinks; with the temperature cooling we thought they may appear to feed etc. Unfortunately, however, the skinks did not surface, but the bumster we thought we had got rid made another appearance, this time without his dog.

Uninvited, he joined us, and began telling tales of seeing snakes (not lizards). I had become totally disheartened with the whole bumster thing, and I just switched off from him. I wondered what was wrong with some of these people; why do they have to stoop to such levels to obtain money? At times, this type of pestering I can only liken to brainwashing - constant and repetitive, they go on and on and on! They have no concept of *"No, thank you"*; they simply refuse to take the hint, and do not leave you alone. All in all, it makes for very difficult working conditions. In fact, rather impossible working conditions, I should say.

I think we left Gunjur Beach slightly downhearted; we had not been hoping to find the little critters, as that would have been too much to ask, but it would have been nice to have been able to conduct some work unhindered. However, to be fair, we now know that Armitage's skink is not extinct; although unsure of numbers, and breeding capacity, from the information we obtained from Luc, it is a safe bet to say it is doing OK. Richard also tells me that the habitat range has increased by some thirty odd miles. So, maybe the team should not feel so down after all, as we have achieved some results today. However, upon returning to the hotel this evening, we were all but besieged by locals who had heard the radio broadcast that Richard had done earlier that morning. They *all* knew where to find Ninki Nanka, but for a fee of course. I was glad that the interview had been done in the latter part of the second week of the expedition; if it had been done any earlier I am sure I would have compromised our expedition efforts.

July 14th

Today was my birthday, and I arose early in order to ring England, as my mother had told me that my son wanted to sing `Happy Birthday` to me. Gambia is only an hour behind, so I made the call at breakfast. It was lovely to hear his sweet voice sing to me, and as soon as he had finished, he asked me if I had found the 'monster' and if I seen any snakes. Before I had a chance to reply he then went into great detail with his latest exploits in his new X-Box game. (I thought to myself, `well he's not missed me *that* much!`).

We had arranged to do a number of things today, the first of which was to dig up part of the area around *Destiny's* nightclub, and the second was to wander around the Palma Rima Hotel area today, as there was a story that a Ninki Nanka had lived in a hole in this area. Apparently, when the hotel was being constructed, it caused the creature to venture from its dwellings. The locals, in a bid to stop it returning, planted a baobab tree. Therefore, our mission was to find that tree.

Yet again, while making our way to the hotel, we were descended upon by bumsters of various dubious descriptions; one even tried to make out he worked at our hotel, and then he tried the con of knowing our taxi driver. I observed Richard deal with this man, and began to smile as I could hear his tones of: *'Oh do you now', Oh really';* At last, I thought: He's got the idea.

The hotel was closed, and this hindered our chances of finding a possible witness to the events of some seventeen years ago, so we began to walk around the perimeter, and look for a baobab tree that may fit the description of the story. We did find a large baobab tree, but I had serious doubts as to whether it was the one from the story, as it was some distance away from the hotel, and it looked far older than the age of the story. However, on closer inspection, the tree did seem to have some significance as there seemed to be many pieces of paper rolled-up, and pushed inside various purposefully and naturally made holes in the bark of this huge tree. This is quite understandable, as the baobab tree, or monkey-bread tree, as it is also commonly known, is a very remarkable tree, and it is quite easy to understand why it is considered sacred and revered by most in Africa.

The tree itself looks as if it has grown upside down. Many baobab trees have an enormous trunk; some have been known to measure 130 feet in circumference. It bears its fruit in December, which are about twelve inches long; they are gourd-like fruits called monkey-breads (hence the name). The husk of these fruits is hard, but the inner flesh is juicy and slightly acidic, but very tasty by all accounts. More than thirty different uses have been found for this tree; the green leaves are rich in vitamins, and can be cooked like spinach, and the young shoots are likened to asparagus. The kernels contain fifteen per cent oil (baobab oil can be made into soap), and can be eaten roasted or raw. The flesh of the fruit can be crushed and mixed with water or milk to make a nutritious drink, while the burning of the fruit produces smoke, which acts as a great deterrent for irritating insects. The husks of the fruit have many functions, such as fuel, serving dishes, and storage. Even the flowers can be eaten raw, or even used to make glue.

The bark from this amazing multi-functional tree, can be used for tanning leather, and the fibres used in the production of rope, fishing twine, baskets, mats, and various other textiles. Its fresh wood contains a great deal of water, and in dryer times people and animals chew it in order to obtain moisture. The baobab is also used a great deal in folk medicine. All in all, it's quite understandable why prayers are offered to the baobab, and that it is considered sacred. The tree is also linked to the Ninki Nanka, although the folklore is somewhat vague, but it goes along these lines:

When young, Ninki Nanka, are said to live in the upper branches of the tree, and remain there until a time when they are too large, and the tree can not sustain them. I can see the parallels with monitor lizards here, along with those of some snakes. Over the years, the folklore, stories, and fables, of such creatures may have become entwined, with a modern interpretation added, and thus we may have ended up with the Ninki Nanka of Palma Rima, as dragon, I have come to know, in these parts may mean snake or lizard.

We moved on to Baka's shop, but he was not there as it was Friday, which is a holy day for Muslims. Suzi and I bought a number of items from the stall, before moving on. The guys had been looking more closely at the maps, and had come to the conclusion that, if there were any, the remains of 'Gambo' were situated just outside the walls of the nightclub. As we made our way to the spot, we amassed a crowd of hangers-on and bumsters. We had to do some quick thinking, and came up with the idea of appearing to be geologists who had come to survey the beach in order to measure erosion, as we were concerned that it may erode as badly and as quickly as the beach at Baku.

This was a fantastic story, and they all fell for it, and it felt so marvellous to get our own back on the bumsters. I sat next to Dr. Chris as he produced his GPS gear, and began to speak into his watch. In order to hide my smile of amusement, I bent my head and made out I was taking notes from what he was saying; the subterfuge was working perfectly. Chris M got really into it, and took the crowd of bumsters off for a lesson in ecology, which enabled us to dig a test hole with our plastic spades. Now, being an archaeology student, I was horrified at the method now being applied; random digging, but a desperate situation requires desperate measures, and I offered to photograph and film the whole thing, as while doing this, I could also keep an eye out for any official looking individuals.

We were able to dig two fairly deep pits. It was very quickly apparent that, at not too great a depth, the sand became very moist and damp, and if there were any animal remains buried here, they would have not lasted too long, and most definitely not twenty three years. I could not help thinking also, that there was also a great deal of building rubble from the construction of the nightclub. We re-filled the holes, and ended our little unique excavation, by telling the locals that their beach was safe from erosion, at which news they seemed genuinely relieved; after all, it could have had a devastating effect on the bumster's economy. We all sampled a fresh mango juice from a beach vendor and basked in our own cleverness at finally getting one over on the bumsters.

We did also manage to get some interesting news on the beach-front, to the effect that, for some years now, anything that is washed-up on the beach is removed by their local council, so as not to impede on the tourist trade. It made me wonder what else may have been washed-up, and been discreetly disposed of.

We got back to the hotel, and the whole area, both inside and out seemed to be buzzing with what we were doing and we got out of our cabs to locals shouting *'Ninki*

Nanka hunters' and *'I know where you can find it'*.

As we entered the hotel reception, Richard was taken aside by the duty manager who had heard his BBC interview and told him that he had some information. He said that when he was a boy (late 60's or early 70's) he had heard a radio broadcast advising people to stay away from an area called Fajara. Apparently a priest by the title of Bishop Maloney, had seen a 'dragon' crawl out of a rocky creek area of standing water, and enter the sea, leaving a large furrow-like track behind. He also said he once knew an old man and his friend, who had seen the Ninki Nanka; the friend had become ill, his hair fell out, and he then died, although the old man did not suffer any effects from the encounter.

July 15th

This turned out not be such an eventful day, more of tying-up loose ends I think. We managed to track down Sueliman, the security guard, and interview him regarding the information he had relayed to me some days earlier about the Ninki Nanka. I filmed the interview, while Richard asked all the usual questions. Sueliman mentioned most things that others had, but he was the only person to mention that to one side, the crest had some form of text on it. Sueliman believed that this dragon's crest was supposed to have words from the Qur'an written on it, and if these words were read you would die. This text may be some form of patternation, as Arabic is a very swirling fluid text, or it's a possibility that in the past, maybe the start of Islam, there may have been some form of worshipping of an unknown creature, and to stop this and to facilitate the worship of Islam, this may have been added to the folklore. Who knows?

Sueliman said he knew where there was a Ninki-Nanka, but it was some distance away and we did not have the time, or the resources, to investigate now.

July 16th

Yesterday, shortly after breakfast, Richard got a 'phone call from a woman by the name of Mariama (who works at Abuko National Park); saying that a man had told her that he knew where Ninki Nanka lived. He said it was in a hole in some mangrove swamps, in a place called Mandinari, just down the road from Abuko, and that he also said that if you threw a dog into the hole, it would bring Ninki Nanka out to eat the dog. Richard was interested by such news, but point blank declined the use of a dog, opting for a chicken instead.

We had already decided to go to Abuko in search of the English guy who was said to have seen Ninki Nanka some years ago, but all we had to go on was his Mandinka name of 'Sudokodo' (we were hoping to be able to check with the records kept at the lodge house on site), so calling into Mandinari was no problem really as it was en-route.

Just as we were all sitting down to dinner that evening, Richard was summoned to the reception, as there was a 'phone call for him, so he hurriedly left the table. Some time passed, and he returned with a somewhat stern look on his face. Apparently, the call was from the actual man who allegedly knew where Ninki Nanka could be found. He had asked Richard if he was the man that had been on the wireless the day before, 'the Ninki Nanka hunter', to which Richard had replied in the affirmative, and the man had then said he that knew where there was a real Ninki Nanka at Mandinari. He could lead us there, but he would not go all the way, as he was fearful of death if he did so. He said he would go within 50 yards of it, and point the rest of the way. He then said, rather ardently, that if you see this dragon 'YOU WILL DIE'.

Richard replied that he was not scared of dragons, and that we have many dragon stories in England. The man then said that this was different, as this was an African dragon, and then proceeded to ask repeatedly *"what do I get?"* whilst telling Rich to ask the people in reception to *"shut up"* as they were making a noise. By this time Rich's suspicions were raised, suspicions which turned out to be well founded when the guy started to make demands of upwards of £2000, £3000, £4000, £5000, as if in some cryptozoological auction!. Rich politely told the man that he would have to discuss this with the rest of the group, and it was agreed by all to decline the offer. However, we did elect to go to the mangroves at Mandinari, en-route to Abuko and have a look round ourselves.

We woke Sunday morning to heavy humidity, the worst it's been during our time here, which was surprising as it had rained heavily during the night. We had breakfast, got ourselves ready, and by 10.00 am we were waiting for our trusty chauffeur, guide and protector, Assan. We climbed aboard the mini-bus and asked him to take us to Mandinari. The journey was somewhat bumpy, as the downpours during the night had made an effect on the road, namely that any infill had been washed away leaving large pot-holes strewn across our route, but Assan, with his ever careful driving, avoided most of them.

When we reached the mangrove area of Mandinari the tide was out, so we were able to negotiate into the swamps a little way, crunching along the shell-strewn mangrove mud and getting a feel of the environment. There was a small wooden and somewhat rickety jetty that some of us ventured onto, at the end of which were the local village children swimming. The only wildlife we observed in this vast swampland were little bee-eaters, mudskippers, fiddler crabs, and the locals who began to follow us and ask us for our possessions; at one stage this included Suzi .

Having only wandered and explored a small proportion of this mangrove, which as a total mass is approximately 10km, it is quite easy to see how people can become disorientated and lost, and why such places hold great mystique and allure for humans ascribing it as the home to such creatures as the Ninki Nanka. We bid farewell to the small group that had become interested in us, and made our way to Abuko.

On arrival, we asked if anyone could remember Sudokodo, but they didn't, so we made our way into the park, and towards the Darwin Centre Tower in a bid to observe the Nile crocodiles they have there. We watched for some time, but all the crocodiles would give us was a few bubbles. We were then joined by a not-so-expert-guide, who led us to the other pool where we actually saw, in total, eleven baby crocodiles so it's good to know that they are breeding successfully. We then made our way round Abuko; unfortunately, the guide we had was not very good and made more noise than all of us put together, so on this occasion we did not see a great deal, apart from a ground squirrel, vervet and red colobus monkeys. I also noticed that Chris Moiser was, in fact, giving the tour guide more information than he was giving us. We reached the café and stopped for some well-needed refreshment, and as we departed the guide then tried to tell Chris that the Aldabran giant tortoise [1] was native to the area! Chris corrected him on this matter and we moved on, but as we neared the exit this so called guide began working his way down our line asking for, well anything really, money, watches; he was most unprofessional, unlike the first guide, Musa Jatta, who really did know his stuff. Ahh... well the luck of the draw I suppose!

On leaving the park, we made our way across the road to Assan who was waiting for us in a little bar owned by a Rasta by the name of Max, who was a refreshingly honest chap. Max told us of a place he thought the Ninki Nanka may reside,' "Bintang Bolon is a place where you may find what you are looking for", he told us. However, he wouldn't go because he had a near-death experience when he had nearly drowned there as a small boy. Even now he will not wash in the river. He also told us that the founder of Abuko, a Mr. Edward Brewer, had at some stage erected a mirror akin to the one that was erected in the 1940s after Papa Jinda was supposedly killed by the Ninki Nanka. It is not clear whether this was done to appease the locally employed people at the park, or for other reasons.

As we approach the end of this expedition, I feel we have more new information and leads than answers to the questions we arrived with surrounding Ninki Nanka, and I strongly feel that the possible root to these answers lies in less well explored areas of West Africa; a possible candidate for this I feel would be Guinea, as it's relatively tourist free and poorly explored. All in all it's been one heck of an eye-opening, interesting, expedition.

July 17th

Chris M suggested that we should pay a visit to the local offices of the *Daily Observer,* the local paper. The plan was to tell them of our story in the hope they would let us look through their archives, in the hope of finding any information regarding Ninki-Nanka. The offices were what I expected - along with the rest of Gambia they were worn, and had seen better days. We realised quickly that they had no archive office here, and so we, or rather Chris M, explained what we were doing here. This

---

1. The **Aldabra Giant Tortoise** (*Geochelone gigantea*), from the islands of the Aldabra Atoll in the Seychelles, is one of the largest tortoises in the world. Similar in size to the famous Galapagos Giant Tortoise, its carapace averages 120 cm (47 in) in length. The average weight of a male is around 250 kg (551 pounds), but one male at the Fort Worth Zoological Park weighs over 360 kg (793 pounds).

did generate some interest, and one of the main editors came into the room and asked if we had heard the radio interview yesterday. It was explained to him that Richard was the guy he had heard on that programme and we were ushered down the road to an internet café where we uploaded the story from Suzi's memory stick on to that of the reporter. Not surprisingly, we had to pay for the internet café time. We returned to the run-down office and waited for some time in what I think was a communal work space; however there was no work going on as a young woman was fast asleep at her desk, obviously the stress of the boredom of the office day had been too much for her!

Then an elderly man appeared from a room with a digital camera and led us out to the veranda to take our picture. Chris M told the reporter that I liked mangos, and he picked one for me - was this in lieu of payment for the story I wondered?

July 18th

We leave for England today, and I must say I can't wait to get into a bath up to my neck in hot water and bubbles, for the first time in two weeks, then climb into my comfy bed and sleep for an eternity.

We stood outside the hotel waiting to climb aboard the coach; even then, I got pestered by the coach driver for money.

The airport was its usual higgledy-piggledy organised disorder; we checked our luggage in, and made our way through security. One of the Custom's security officers I noticed was staring at me very ardently, and I began to think I was in for some routine searching ,when he suddenly pointed right in my face and said: *"You lady. You Ninki Nanka lady in paper"*.

I was filled with relief at the thought of not being subjected to a search.

"Yes I am", I replied.

"Did you find it? Did you see it? You know you will die if you do?"

"No, we didn't," and I began to point out the rest of the team who were behind me at the X-ray machine in order to make a quick exit to the waiting lounge on the other side. I thought to myself: Who would have thought that my 'Warhol, fifteen minutes of fame' would have been in all places The Gambia?

END NOTES

As the plane took off and banked round and away from the coast of Gambia, I could see the river twisting, turning, and shimmering, almost snake-like in its own appearance. I began to ponder over the information we had amassed over the last two weeks, and in an attempt to make sense of it all, various aspects kept coming to mind:

1. There was Baka Samba whose relation had seen this huge creature complete with fire emanating from its mouth.

2. The guide from Abuko Musa, who believed it to be like a huge python, but with legs and wings like a bat, and who also thought it spewed fire from its mouth.

3. The grandson of Papa Jinda, Hassan, who believed it to have a crown of fire atop of its head, with its body covered in rainbow glittery scales.

4. The Islamic text to one side of the crest/comb.

On the plane, Richard told me the information he gained from that day at the fish-market in Baku:

It's very snake-like but can grow to the size of a palm tree, its scales are very shiny to the point of being mirror-like, and when it gets to a certain size ,it transfers itself to the sea.

It made for quite an eclectic mix of creature: long huge snake-like body, huge shimmering scales, a crest/comb that sat atop of its head with a high probability of death if you saw it. These were about the only consistent attributes that kept cropping up; I could not help wondering if there was, in reality, at least two or a number of separate unknown creatures being described here. If it exists at all, it's a possibility that can't be ruled out.

There is one other thing; most of the accounts have been handed down through generations, from elder members of families or communities, stories passed down through time from grandparents, uncles that were hunters, in fact from a bygone age when Africa was very different and indeed, the 'Dark Continent'. Perhaps in a long forgotten time, there were such mysterious creatures that may have been revered, and worshipped, but with the advent of Islam, and possible extinction, only the odd folklore tale has survived. These oral traditions have then been passed on, and given a modern interpretation, as they have been passed down to the present day. These modern additions, to very possibly ancient tales, serve to keep the Ninki Nanka very much alive in the minds of the Gambian.

We have jointly come to the end of another CFZ expedition, and I feel we have more new information and leads than answers to the questions we arrived with, surrounding the mysterious Ninki Nanka. I strongly feel that this creature is not native to Gambia, but travels through it on occasion. I feel its home, although still in West Africa, lies in less well explored divisions. A strong candidate for this would be Guinea, as it is relatively tourist-free, as well as being very under-explored, and, as a consequence, very little is known of its fauna.

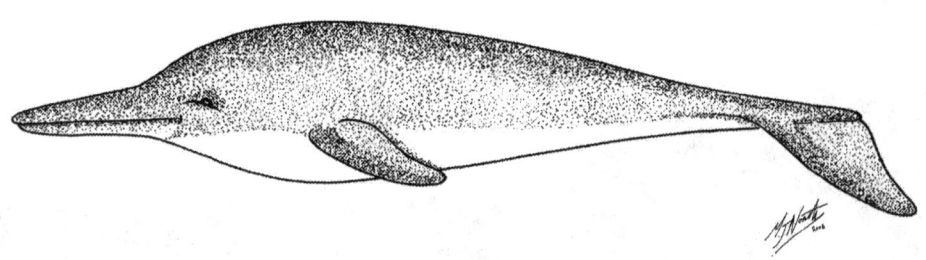

**Did 'Gambo' look like this?
Based on the descriptions provided by Baka**

My first taste of Africa has found it to be a place teeming with life, and in the most part, a joyful people. It is a place that is very slowly making its way towards democratisation, despite corruption being at high levels in government; sadly this corruption filters down into the everyday lives of Gambians, but to be fair these same governments are beginning to realise what an abundance of riches they have in their flora and fauna. The revenue from tourism is certainly a huge motivating factor for the conservation of landscapes such as the mangroves of Gambia, and with more education, the Gambians are becoming more aware that, instead of thinking in the short term, they need to apply long-term strategies and views, as they begin to realise that their economy depends on intact ecosystems such as the mangroves, dry forests, and coastal shoreline, of Gambia.

Not only was this my first taste of Africa, it was also my first jaunt abroad with the CFZ, and I have enjoyed being part of the team immensely. However, at times, I have become somewhat confused and disparaged with the descriptions and alleged sightings of Ninki Nanka, even more so after the BBC radio broadcast, when a minority of individuals seemed to be aware of the whereabouts of what they perceived to be a Ninki Nanka. However, they were only prepared to point us in the general direction of one (for a ridiculous fee, naturally) as the fear of seeing it, even in the 21st Century, seems to evoke waves of fear within certain groups of local people.

As for `Gambo`, only the sands on Bungalow Beach have the answer to that question, and with so much development occurring in the last 23 years, the beach is almost completely different with the advancement of buildings etc. My money would be on the actual creature in all probability being a rare type of dolphin.

But amongst all this uncertainty it's good to know that the little Armitage's skink is now known not to be extinct, but in fact is alive and well, all be it in toilets of the bars of Gunjur Beach.

The headland outside the hotel

Headland by day

The ruins of the pumphouse at Abuko

Vervet monkey (*Chlorocebus* sp.) at Abuko

Hooded vultures

Five lined skink *(Euprepis margaretifera)*.

The sign at the Ninki Nanka cafe in Abuko

Walking the trail in Abuko

Baka's Ninki Nanka pendent

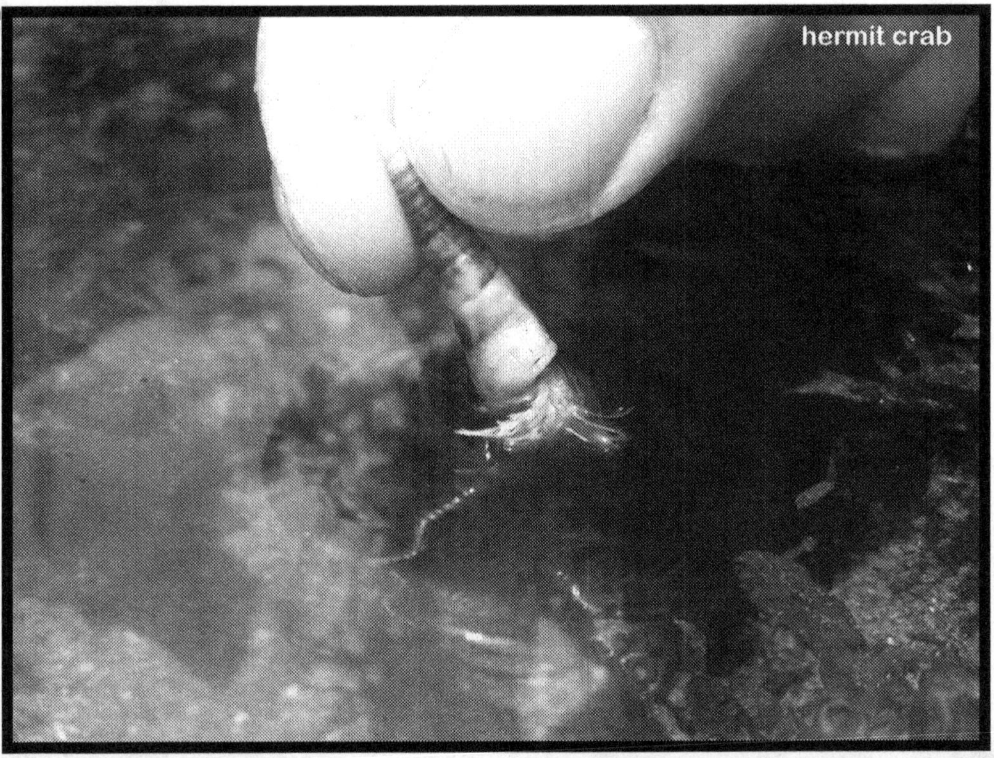

hermit crab

Lisa with a cuttlefish

Richard with the cuttlefish

women sell their wares at the market

Drying catfish bladders for export to China

fisherman with net

Chris Moiser with Assan the taxi driver

The front of *Destiny's* night club

The internet cafe where we posted our blogs

Babu: he kind of helped us out

Rainbow agama

The view from our verandah

Prayers inserted into the baobab tree

Boats on the Alahein

Alahein River

Pool in Senegal

Mangroves in Senegal

A juju stick embedded in a pile of empty oyster shells to ward off thieves

Lantana flowers

Young nile crocodile

Mole viper (*Atractaspis sp.*)

The sand dunes of Gunjur

Lisa picks a succulent jungle mango

Richard talks about vestigial legs on pythons

Oll Lewis

Oll Lewis is the ecological director of the Centre for Fortean Zoology (CFZ) and has studied and investigated many cryptozoological reports in his native Wales, his achievements including photographing and unmasking the Penarth mystery cat as a giant domestic cat. He is the leading authority on Welsh lake and river monsters and also heads the CFZ's aquatic monster study group.

Oll currently lives in the Centre for Fortean Zoology based in Woolfardisworthy, North Devon where he manages the CFZ archive and is currently designing the

centre's museum (planned to open in 2007) with the director Jonathan Downes, which will be the United Kingdom's only museum dedicated to Cryptozoology as a whole. At 26 he is one of Britain's most prolific young cryptozoologists and has an honours degree in applied biology.

Day 1, 4th July - the arrival

We arrived in Banjul airport in the mid-afternoon, and disembarked from the plane into the hot Gambian sunshine, to see vultures wheeling in the skies above the runway tarmac. Hopefully, this was not an omen in a similar vein to the 'corpse birds' that are said to fly around Welsh pit-heads, to signify an upcoming disaster, or the deaths of miners.

Perhaps, however, if it were any kind of omen, it was of the 'porters' that lurk in the shadows of the luggage reclamation hall of the airport, venturing out of their shadowy eyries, equipped with trolleys to scavenge off unsuspecting tourists.

Thankfully, Chris Moiser, who, having been to Gambia several times previously, with large groups of students, knows what to watch out for, forewarned us of these colourful characters. To be fair to the baggage porters, if one has more than one bag, they *do* come in handy to negotiate the scrum of customs, as many of our fellow passengers had decided to become very un-British all of a sudden, and push through, rather than queuing politely, and came cheap at only a pound or two.

The scenery of the Gambia is beautiful: green trees, terracotta-red earth, one-storey dwellings, and rivers, all seem to intermingle with a strange symmetry. The animal inhabitants of the Gambia are a surprise too, and in the few hours I've been here, I've seen a red eared agama up close, five-lined skinks, yellow and white swallow tail butterflies, a speckled pigeon and hooded crows, all of which are fairly common animals here, but to a biologist, used to European species, it's a fascinating experience seeing such a different ecosystem up close.

Tomorrow, we start our investigations into Gambia's stranger and rarer animal inhabitants - things are about to get a lot more interesting.

Day 2, 5th July - Bungalow Beach reconnaissance

Wide-eyed with enthusiasm after a hearty breakfast, we set off to walk through the streets of Bakau to Bungalow Beach - the reputed final resting place of `Gambo`. A young boy and amateur naturalist called Owen Burnham, who witnessed a strange creature being washed up and buried on Bungalow Beach, brought the creature known as Gambo to the attention of the Western world. The creature he described was similar to a pliosaur and, hoping to return at a later date to dig it up, Burnham made a map of where the staff of the Bungalow Beach Hotel had buried the animal in a three-foot deep shallow grave above the tide line. We had that map.

Bakau is an interesting place for a walk. However, being a portly gent, I hadn't quite adapted to the humidity of the Gambian rainy season yet, so was proceeding to melt in the hot weather, and had to encourage the group to take regular rest-stops in the shade along the way. It was at one of these rests, that we found ourselves across the road from an imposing, but picturesque, looking villa with high exterior walls. A man, probably a secret policeman, or the Gambian equivalent thereof, approached us to enquire politely as to what we were up to. It was only after, that we realised that the villa was the current home of the deposed ex-president of the Gambia, so our choice of rest stop had probably worried the security personnel somewhat, especially after the recent attempted coup in Burkina Faso involving Mark Thatcher, and a recently foiled counter coup in the Gambia.

Along the roadside, we saw several interesting invertebrates, the most impressive of which being large red velvet mites and giant millipedes.

Halfway through the journey, we ventured on to the beach where we were joined by unwanted guests. According to a new friend of ours, called Lamin, who we would meet later in the week, and who co-owned an internet café, the seventh guest, 'Glue', was one of a group of people the locals called bumsters. Bumsters are vagabonds that will pretend to be tourists' friends, or offer themselves as guides, in the hope of money, or even better, a free passage to Europe and its riches, by befriending a wealthy middle-aged widow who wants a new young African lover. Europe's riches are exaggerated, as all we have is a very favourable exchange rate compared to that of many African countries, but as we also get most of our income squirrelled away by taxes, I'd imagine that if bumsters ever do make it to Europe in the tow of a wealthy widow, they are in for a bit of a shock.

Eventually, we made it to the last resting place of `Gambo`.

However, we had problems. Since Chris Moiser's last visit to the Gambia, Bungalow Beach has had a bit of building work. The patio of *Destiny's* nightclub, currently owned by the president's brother, had been built dangerously close to where the 'X' was on our map. It may even have compromised the burial site, but that would all depend on how deep the foundations of Destiny's went and how far they extended.

With our initial optimism tempered by the existence of the nightclub, we made our way to the nearby market to see if we could procure some pendants that a local jeweller was known to make of the Ninki Nanka. Also, markets are good places to find information, and it was possible that we may find a witness, or a good lead, that would help us, either in our search for the Ninki Nanka, or `Gambo`. As it turned out, the man who owned the store next to the jeweller, Babu, aka Mr Fixit, could help us on both counts. As we were buying the pendants from the jeweller, we asked if he had ever seen the Ninki Nanka, but he hadn't. However, Babu told us that his uncle *had,* and even better, he had personally witnessed the beaching and burial of `Gambo`. This was fantastic news, because other than Owen Burnham and his family, there has never been another documented eyewitness of the `Gambo` event and no-one to confirm Owen's story. [1]

---

1. Having interviewed Owen on a number of occasions, I am convinced of his veracity, but, until now, the story has been essentially uncorroborated. JD

In order to confirm that he was talking about the same event, as Richard and I interviewed him, I tried to get Mr Fixit to draw a picture of the animal he had seen. Unfortunately, he said he couldn't draw, so I had to draw the animal he had seen according to his instructions. He fished out a pendant of a dolphin from his stand, and said it looked like that (in fact he had insisted it was a large, unwell, dolphin from the start of the interview). Something I remembered from Owen's account was the missing - fins of `Gambo`, so as I was sketching from the pendant, I asked Mr Fixit if particular fins were there, and the positions of these. He was very insistent that there was no dorsal fin, but whether this was removed by trauma, or if it was stunted, or never present was not clear, but this certainly tallied with Owen's testimony.

Day 3 , 6th July - Abuku National Park

Today, we decided to visit Abuko National Park, as it was the setting for one of the most well-known sightings of the Ninki Nanka, and also so we could learn more about the ecology of the Gambia in general. We left our base at 9.00 am in a ten-seater taxi driven by Chris Moiser's long term associate Assan, and after a 45 minute journey through pot-holed tarmac roads, and compacted mud streets, we arrived at the National Park. The entrance fees were very reasonable, coming to about 50p each. As soon as you pass the perimeter fence in Abuko, you enter the jungle, and start to see animals; almost immediately we spotted a giant millipede walking across the path, and when we reached the first hide we were able to watch crocodiles in the wild. Richard, in particular, was overjoyed by this, as it was, despite his having been reptile keeper at Twycross Zoo, the first time he had been able to see one of his favorite animals in the wild.

It was by the first hide that we met a guide called Musa Jatta. After being followed by a so called 'guide' against our will yesterday, as we went on our recce to Bungalow Beach, we were wary. We needn't have been, however, as Musa Jatta proved himself to be a first -rate spotter of animals (even better than me), and, perhaps, even saved Richard from getting on the wrong end of a spitting cobra too. Whilst all of this was happening, Chris Moiser had walked on ahead and was talking to a lady with a carving shop at the half-way point of the path circuit around the reserve, from whom he found out more about the famous sighting in the reserve's area. She also told him that the grandson of Papa Jinda, the man who had seen the Ninki Nanka, was a young man called Hassan, who actually *worked* at the reserve.

We all met up by a small drinks stand called the Ninki Nanka café, and Richard asked our guide if he had heard anything of the café's namesake. Musa had. There was, apparently, a report in August, a few years ago (we pinned it down to 2001-2003), when the beast had apparently left a long furrow in the mud road, and the driver and his passenger were said to have died when their lorry ran into the furrow during heavy rain. Musa's description of the animal was: *"like a huge python, big enough to swallow a cow"*. It had legs and bat-like wings and could spit fire. He did not know if it could fly, but it could move in water and over land

As we made our way back to the entrance, we passed the remains of the pumping station that Papa Jinda had been working at when he saw the Ninki Nanka. The whole area, with long established trees growing through the derelict buildings, rusted machinery, and long abandoned dams looked like a cross between the temple of Ankor-Wat and a level in Tomb Raider. When we made it back to the entrance, spotting a brightly patterned juvenile monitor lizard *en route*, we found Hassan, and were able to interview him about his grandfather's sightings of the Ninki Nanka.

Papa Jinda had witnessed a scene of devastation at the pumping station in Abuko, when a Ninki Nanka had destroyed several pipes at the pumping station. The mention of a Ninki Nanka had caused a panic among the workers, and they had asked for a mirror, as it was commonly thought that the only way to get rid of the animal was to show the beast its own reflection. The second time Papa Jinda came into contact with the Ninki Nanka, was to prove fatal, as after seeing the beast he fell ill, complaining about pains in his legs and waist, with his hair also falling out. He died two weeks later. The Ninki Nanka, being seen as an omen of imminent death, either sudden, or within four years of seeing it, is one of the few aspects of the folklore surrounding it that has been consistent in every case we've had reported to us so far. If we do find the creature on this expedition, we can only hope that this will prove to be incorrect. Hassan said that his grandfather had described the animal as having what looked like a tiara of flames on top of its head, and it is a possibility that this was a red and orange crest.

When we asked about recent sightings of the Ninki Nanka, Hassan told us of a man who had found what he claimed were scales of a Ninki Nanka nearby. This was too good an opportunity to pass up, so whilst Richard and Lisa went back into the jungle, accompanied by Musa on a bike to try to find the man, Dr. Chris Clark, Chris Moiser, Suzi, and I, hopped into Assan's minibus, accompanied by a policeman, on a mission to find the man with the Ninki Nanka scales. After tearing around the local streets in a manner that would put local TV show *Banjul Cops* to shame, we eventually pulled up outside a local slaughterhouse, where our quarry had just finished prayers for that part of the day. A look of fear passed across the wizened old man's face, as the burly policeman shouted at him to *"Get in!"*, but he soon brightened up when he realised he wasn't in trouble.

Eventually, we were shown the scales of the Ninki Nanka, but sadly, from my examination of them, I think they may just be a bit of rotted film cell. Richard and the others are certain that they are not biological, least of all scales or skin. However, we have acquired a sample to test when we get back to the UK, as it would certainly be bad science not to investigate every claim as thoroughly as possible.

Lastly, before we left, we met up with another employee of Abuko National Park, Bakari Jarju, who claims that the Ninki Nanka has been seen recently in a lake near a village 80km away. Richard, Dr. Chris, Suzi and I, plan to investigate this on Saturday.

Day 4, 7th July - Preparations to go up river.

As we had a very early start tomorrow, it was thought best if we had an early night tonight. As we weren't sure whether our guide would be trustworthy, or was just looking for a cheap way to visit his relatives, Assan, our taxi driver, came to our hotel later that evening with the four-wheel drive we would need to take, for our perusal, and its driver, a young man called Kamera. Assan assured us that if there were any problems with the situation that they would get us out.

Day 5, 8th - July Up River

We woke up bright and early at 5.00 am, picked up the packed breakfast the hotel had provided us with, finished our packing, and went out to meet Assan, Kamera, and the open top four-wheel drive we had become acquainted with yesterday.

Driving long distance in the Gambia is an interesting affair as, outside of the cities of Banjul and Bakau, the roads steadily get worse, starting with the strong tarmac being replaced by a mix of tarmac and shells (which is less hardy and leaves itself open to extreme pot-holing whenever a shell is eroded or winkled out of the road surface), and, further down the road, the whole idea of tarmac is abandoned altogether.

When the tarmac goes, you are left with two-lane mud tracks that are effectively one lane, because the erosion of the road is so bad. The driver is forced to weave the car between potholes so big they would more accurately be called 'potpits'. A canny driver will often drive in the road's wide gutters, which are usually in better condition than the roads themselves.

The other thing that makes driving 'interesting' in the Gambia, is the intermittent police and army checkpoints. Rather than drive around as traffic patrols do in the UK and Europe, the police set up roadblocks on the main roads, to check driver's documentation.

The cumulative effect of these factors on driving, made the 80km drive for Richard, Dr. Chris, Suzi, Assan, Kamera, Bakari (who we picked up en route by the entrance of Abuko) and I, take four to five hours.

En route, we stopped to investigate the scene of the accident that, according to Musa at Abuko, was reported in the newspapers as having been caused by the Ninki Nanka. A truck was said to have been driven off the road when the Ninki Nanka crossed it, creating a massive ditch, with the accident killing the driver and his friend in the lorry's cab. According to Musa, the report had appeared in the papers about three years ago, and the wreckage was still there, left exactly as it was after the accident by the side of the road - it had to be worth a look.

We disembarked from Kamera's vehicle to check out the drainage culvert that allegedly had been used to repair the road where the ditch had been. Looking closer at the culvert, Dr. Chris was first to voice concerns of the veracity of the story, because the culvert looked weathered, and very eroded; too badly eroded to have been there for only three years. The culvert was not the only thing that didn't add up about the story, and what was at the scene. The upturned lorry was some distance down the road from the culvert, and the lorry's cab was so badly rusted it was merely a thin, brittle, dark brown shell; the engine, fixtures, and fittings, having been cannibalised for the spare-parts market long ago - again, from the state of the cab, much longer ago than three years. Next, I ventured into the upturned trailer of the lorry. The trailer had been open-topped so this had allowed new growth under the trailer (which had two 'windows' in the side that allowed light in). As well as the new growth of plants under the trailer, two trees had grown, along with the contours of the bottom of the trailer, which was, of course, now the roof. These trees were not bendy saplings, but wooded trees of at least eight years old. All this meant that, even if a Ninki Nanka had caused the accident, and not the road just collapsing due to heavy rains, as a more prosaic explanation might have it, it had certainly happened longer ago than just three years. Unfortunately, as the *Observer,* the paper in which the story had appeared didn't archive its stories; this would be as far as we could get with investigations of this particular incident.

We made it to Kiang West National Park just before mid-day. The contrast between Kiang and Abuko is stark. There is no entrance fee in Kiang, and also, at least when we were there, no guides and, as far as we could see, no tracks apart from the road. The vegetation and habitats are also very different - Abuko is jungle woodland, thick with trees and shrubbery, Kiang had tree-cover in some parts, but the trees covered the area much more sparsely. We were led by Bakari, and a local he had picked up at the nearby village of Dumbletore, through the undergrowth, and down a steep valley side to a meadow, where long grasses were being grown to make thatch for the local village houses. We walked further in the baking mid-day sun, and eventually came to a seasonal lake dried up by the sun, apart from a few marshy areas. The still, slightly moist, silt of the dried up lakebed, had done a fantastic job of preserving animal tracks, and during this part of the walk we recognised the tracks of a hyena, baboons, and a civet or mongoose, although we kept an eye out for any other more unusual tracks as we were approaching the area a Ninki Nanka is said to inhabit. However, if there were any Ninki Nanka tracks, we certainly didn't see any.

Eventually, after a mile or so walking, we made it to the place we'd come to see. There was a clear difference in the vegetation as the cultivated area came to an end, and the local guide ran to a stand of trees, before the start of the uncultivated area, to hide. Bakari was a bit braver, and walked a little way into the uncultivated area, and translated what the local guide was telling us, as well as letting us interview him, and the local guide, on camera about the local Ninki Nanka beliefs. It was, perhaps, the strangest interview I've ever been party to, with the local man shouting his answers from his tree about 25 metres away in the local language, and Bakari translating. The local man told us that the people from the village would not go beyond this point, as the Ninki Nanka was seen in the vicinity, and it had killed a hunter there. Local belief held, also, that a nearby lodge had had building work disrupted because of the Ninki Nanka, although, according to Bakari, it was just because funding ran out for the project. As we interviewed Bakari and his friend about the nature of the beast, the familiar topic of death came up. Like many other places and people we've asked about the creature, it is a belief in Dumbletore that seeing the dragon will cause your death. However, here there was a particular twist, if you didn't tell anyone about your dragon sighting you would likely survive, but if you were unwary enough to tell anyone about it then you would die soon. The belief is that the Ninki Nanka still holds power over you after you have seen it. I asked Bakari when people were more likely to see the animal, and he revealed that you are just as likely to see it in the day as in the night. According to Bakari, he sees it more as a spirit than a real animal.

We walked down through the uncultivated area, and came upon a pass through a mangrove swamp. The mangrove pass was hard to traverse, and it was clear as we clambered over fallen trees, large igneous rocks and mud, that despite the odd discarded shoe or bottle, which had probably been washed down in floods or when the lake was full, that this was a rarely visited and forgotten place. We did not see anything particularly strange here, and eventually reached the shore of the River Gambia, and walked through bushes along the bank, ending up in the abandoned lodge. After a rest from the blistering heat in a shelter, and a thirst-quenching lunch of mango and paw-paw, we headed back the way we had come.

As I was not a fast walker, I soon found myself lagging behind the group, and heard a loud crashing noise in the bushes behind me. Was it the Ninki Nanka? Sadly not, but perhaps it was a source of the local belief that the Ninki Nanka lived here. I turned around to see a large monitor lizard sheepishly sloping out of the bushes. In terms of the overall size that monitor lizards are known to grow to, it was nothing too unusual for an adult, but there is a practice of culling any monitor lizard seen over a certain size. Large monitors are rare in the Gambia as a result, so it is conceivable that a monitor lizard which managed to grow large by evading the attentions of men with sharp blades, could give rise to local sightings of the Ninki Nanka.

We made it back to the four-wheel drive, and were happy to be greeted by Assan and Kamera with bottles of water and ice cubes from the cool box - after that sun we *really* needed it. As we drove through the village, Bakari attempted to find an eyewitness he knew about, who was a former ranger at Kiang West, and as it appeared that he was near the river, that's where we headed to.

Unfortunately, my shoulders had started to sting from the hot sun, and I was paying the price of using a factor 15 sun cream (as recommended by the chemist back home). My sunburn was beginning to hurt, so I had to sit this out in the car. This gave the others the chance to travel light, and leave the backpacks with me in the car. However, after a few minutes, they came urgently running back, the batteries in the video camera, and Lisa's camera, had run out seemingly concurrently, so they needed my camera to take photos of the witness. I suggested they take Lisa's Dictaphone too, which was languishing in Richard's bag.

Crisis averted, they went back out to interview the eye-witness (Momomodu), who claimed to have seen a 50m long by 1m wide snake, which led to him (Momomodu) getting blistering and flaking skin, that was later cured by a Muslim holy man with a special herbal balm.

Back in the four-wheel drive, Richard used his laminated photos of several animals to ask Momomodu what the creature he'd seen looked most like. Out of all the photos and pictures, he settled on a depiction of a Chinese dragon. He also told Richard of an Englishman, who the locals had called 'Sodokodu', who had been working for the Parks and Wildlife Department and had apparently seen Ninki Nanka. We gave Momomodu a lift back to Dumbletore, and we opted to give him a small unprompted donation. However, another villager, who thought it would be a good idea to try to weasel cash out of the 'tourists' by insisting that we had to pay him to enter Kiang West Park, saw this. True to his word yesterday, Assan sent him packing, with a flea in his ear too. The trip back was uneventful and we made it back to the hotel at about 9.00 pm, tired but upbeat after the most eventful day of the expedition so far.

Day 6, 9th July - Fishing

After the tiring nature of yesterday (unfortunately, the sunburn I acquired went rather deep, and would affect me for most of the rest of the expedition too), we didn't do a lot today. We held a meeting to fill-in Lisa and Chris Moiser of the events of the previous day, and to plan for the rest of the expedition. There were several places and leads we had to follow up, one of which had been provided to us by a Welsh photographer, called Ian, whom we had met in the hotel. Back in the UK, one of the aims we had decided on for the expedition was to find, investigate and, hopefully, film, a rare and endangered species of lizard called Armitage's skink, found only in a very small section of dunes in the Gambia. Ian had seen one in a reptile park near to the village of Folonko (the village also being home to a sacred crocodile pool).

In the afternoon, I set up my rod to try my hand at a spot of sea fishing, although this kind of fishing is not my usual discipline, and I am much better at, and prefer, coarse fishing. Rapidly, I began to get frustrated by the strong waves taking my makeshift bait (the halal equivalent of spam). I would have preferred to use worms, as they could have been threaded onto the hook, and not fallen off as readily as the meat did, but after the adventures that other members of the team had encountered in the local fish-market, I thought that, perhaps, suggesting returning to get some bait might not be such a good idea. After a valiant, yet ultimately fruitless attempt of catching anything, and after getting soaked by waves crashing on the stone jetty, I gave up fishing.

Day 7, 10th July Banjul

As the morning sun rose over Bakau, it once again promised to bring with it another blistering hot day. We decided that we would very much like to go to Banjul, in order to try to obtain local folklore books, and check out the national museum to find out if it contained any interesting information.

We were Driven to the Atlantic Hotel, where we arranged to meet our taxi drivers, Assan and Oman, later in the day, and walked past the cricket ground to the market near the town centre. The market is, at the moment, a shadow of its former glory. Since the original burned down in the 1980s it has never regained its past glories and some parts of it are probably not areas where tourists, or people who look like tourists (like us for example), should dally too long. Richard, Suzi, Dr. Chris, Lisa, and I opted to stay in the shade near the market's entrance, while Chris Moiser risked the trials of the place to obtain a copy of a rock song about the Ninki Nanka. Yes, Chris was on a mission. After he returned from the upper echelons of the market, Chris had someone claiming to be a 'security guard' in tow, who we managed to quickly, lose.

We were unable to locate an English book store; the one Chris Moiser remembered appeared to have closed down, and not even the combined might of Banjul's central police station could help us find it, so we went on to the museum. Here, we were able to buy what would seem to be the only book on Gambian folklore in publication, but which sadly provided little information about the Ninki Nanka. The museum staff did not know anything about the Ninki Nanka, and it was not even mentioned in the exhibitions. However, the museum was rather impressive, in particular in its prehistory department, because you hardly ever see anything about prehistoric Africa in European museums, and is well worth a visit if one finds oneself in Banjul.

We returned to the nearby Atlantic Hotel, the oldest colonial hotel in the city (built in the 80's in keeping with the style of the original, now demolished, which inhabited the site next door), in order to enjoy a Chapman's cocktail and see the bats in the hotel's bird garden. Chapman's cocktails are the non-alcoholic cocktails that West Africa was famous for in colonial times, so it seemed particularly apt that we take this opportunity to toast the memory of J.T. Downes, father of the Centre for Fortean Zoology's director Jonathan Downes. J. T. Downes had been a colonial administrator in West Africa and Hong Kong, as well as an explorer, and was the man whose memory this expedition is dedicated to.

However, the barman insisted that he was unable to comply with our request since he did not have a cocktail shaker. Quite why this mattered when a Chapman's is meant to be stirred and not shaken was unknown, but if the barman would make such an elemental mistake in a place that is meant to be famous for the cocktail, his version of it probably wasn't up to much, and we opted for fizzy fruit cocktails instead for our toast, as this was the closest thing available.

After taking refreshment, we followed a winding path to the hotel's bird garden. One gets the impression that the small wood planted to encourage birds is not high up the list of hotel priorities, firstly because there are hardly any birds there and secondly because, whereas there may not be many birds, thanks to the presence of a stagnant pond, half of the Gambia's mosquito and biting insect population seem to have made it their home. The miasma of mosquitoes munching away at our legs made us aware that we were on the menu so, once we had achieved our objective and seen the fruit bats roosting in the trees, we beat a hasty retreat. The mosquitoes had different ideas and the miasma followed us out too, but they soon felt the cold steel of my walking stick and were sent packing. When we made it to the front of the hotel, the men of Bakau's finest taxi service were waiting to ferry us back to the African Village Hotel.

Day 8 11/7 Thinking about `Gambo`

As it was my birthday, and we were going on another big trip up country tomorrow, we opted to take today to write up and send off blogs back to Jon and Mark in the UK. So today's entry seems to be a good point to summarise what we know about the mystery animal of Bungalow Beach or `Gambo` as it is more colloquially known.

In 1983, teenage amateur naturalist Owen Burnham was visiting Bungalow Beach with his family, when a strange animal washed up on the beach. According to Owen, it was around 15 feet long (4.50m), black with grey under parts and had four flippers or fins, but with no discernable blowhole or dorsal fin. Its head had nostrils and a long snout with a row of 80 sharp teeth. If this description were to prove accurate, then it would have to be not just a new species of large animal, but a whole new genus. In terms of body shape then, it bears some similarities to a pliosaur, which was an aquatic reptile that became extinct with the dinosaurs. However, despite the often-quoted similarity between Burnham's description and that of a pliosaur, the animal was certainly not a pliosaur. For starters, if it were a pliosaur then it would not have dolphin-like skin, and pictures drawn by Burnham of the creature give the head of the creature a different appearance too.

However, Owen is the only witness of the beaching of `Gambo` to give it quite this description - other witnesses - his family members - have claimed it was just a large dolphin that was a little the worse for wear. Baka, the new eyewitness we discovered, did, however, collaborate what Burnham said about the absent dorsal fin and estimated its size very roughly at around 10 feet (3M) which, although this seems quite far off Burnham's 15 feet, it could be accommodated within a generous margin of error, bearing in mind that Baka was relating an event that he witnessed over twenty years ago, and using landmarks to give an approximation of size.

`Gambo` was buried 3 feet deep in the sand of Bungalow Beach by staff from the hotel, very close to the current site of the patio of *Destiny's* Nightclub.

Day 9, 12th July - Canoes, cautionary tales and crocodile worship

Today, we set off in two taxis with Assan and Oman, to Kartong and the Alahein River, which marks the southern border between the Gambia and Senegal. Once there, we took refreshment at an Italian café, run by a Frenchman, where we were offered a tour of the river by two boatmen. Moored outside the café was a large fishing boat that, quite frankly, looked like the bee's knees, so, for only 600 Dalasi (£11.60), to take us and Assan on a small cruise, this seemed like too good a deal to pass up and we agreed. Then, in true comedy style, the boatmen collected their boats.

It turned out that they had two clapped-out looking canoes, and the swish looking fishing boat had been nothing to do with them. Richard, Lisa, Dr. Chris and Suzi took one canoe and I went in the other with Assan, as I wanted to ask him about the story of the Palma Rima Ninki Nanka.

As the story went, when they were building the Palma Rima Hotel near to Bungalow Beach, work stopped as the builders caught wind of a Ninki Nanka inhabiting a hole near the site that would go down to the sea to feed. The creature was apparently thwarted only when a large baobab tree was transplanted into the hole, allowing work to then continue on the hotel once the creature sloped off. Assan told me that it had happened, and the tree was still there, but he said that the animal had just been a large python, not a Ninki Nanka, and, therefore, a case of misidentification. I talked to Assan about the witness in Kiang West, who had seen the giant snake he took to be the Ninki Nanka, and about the discrepancies in a lot of the descriptions, along with the similarities. I commented that it was almost as if it metamorphises into different creatures depending on whoever you talk to, and Assan told me that in his opinion the Ninki Nanka was not in Gambia anymore, but had moved to Guinea. I asked Assan whether it would be a good idea to ask the boatmen if they knew anything about the Ninki Nanka, but he was of the opinion that we shouldn't, as they were not trustworthy, and would just tell us what we wanted to hear.

While we were talking about this, the other boat was beside us with its boatman earwigging our conversation. I would later learn, when Lisa interviewed him, that he just repeated what he had heard Assan and me talking about, so that certainly proved to me that Assan had been right in his assessment of the boatmen, and that we shouldn't necessarily believe everything we hear.

On the bank, we also saw a pile of shells with a strange stick placed in it. I asked Assan whether there was any significance to this, and he told Richard and me that the stick was a juju stick that the man, who had collected the shells, had placed there to curse anyone who took them. The shells are often ground up to a paste used for whitewashing houses.

Next on our agenda today, was a visit to the sacred crocodile pool at Folonko. Crocodile worship in the Gambia was one of the most prolific Animist religions in the Gambia before the coming of Islam and Christianity to the area, and some elements of the crocodile worship have been co-opted into the local version of Islam, in the same way that old pagan holy wells and springs became part of the local version of Catholicism in Britain and Ireland. We left the taxis a little way up the road from the pool, and, as we were on holy ground, we removed our shoes and socks. The holy ground itself was covered with fine, cool, sand that felt pleasant on one's feet, and we walked around the corner to the pool to see an almost, empty pool covered in water hyacinth, with no discernable signs of crocodiles at first glance. However, as we stared at the pool what we thought was a twig shifted slightly, and revealed itself to be a juvenile crocodile of about 1-2 feet long (30 – 60cm).

Circumcisions, both male and female, still occur at the pool in the same manner that they have done for centuries, and bathing in the waters of the pool is said to increase fertility. When a person is circumcised at the pool, they are not allowed to bathe for nearly two weeks afterward, before they take their first ritual bathing in the river. When the bathing occurs, a song is sung about the Ninki Nanka to ward off the creature, which a young local man sang for us. It is a tempting assumption to make that, as this practice is the oldest tradition associated with the Ninki Nanka that we have uncovered, West Africa's old Animist religions might be where, at least, the folkloric elements of the Ninki Nanka originate from.

Next, we went to the reptile park, which thanks to the sunburn from hell I acquired in Kiang West, which had been getting worse all day, I had to sit out in the shade. This was a shame, because I really wanted to visit it, but if I got any more sunburn I could have kissed goodbye my chances of recovering before the end of the expedition. It was here though, that Suzi, Dr. Chris, Lisa and Richard met Luc Paziand, and he told them of the Armitage's skinks Ian had seen and photographed, and where they were found. However, one had died, and the survivor had been released only a few days before our arrival. They had been found well outside their known range, in the dunes near Gunja by Rastafarian bar owners who had been digging a new toilet. This, of course, was a lead that we would be foolish not to follow up.

On the way back, one last stop was made, at the house of a friend of Assan's from Guinea, whose grandfather claimed to have seen the Ninki Nanka. I was asleep in the taxi when we stopped though, and by the time I had got my stuff together, the others had disappeared down an alleyway and I was unable to follow. From what Richard told me, the man had a scale like the one we picked up in Abuko and offered to summon Ninki Nanka - it would seem that Tony 'Doc' Shiels has a Gambian protégée.

Day 10 ,13th July - Rastas, skinks and bums

After a day spent making several abortive attempts to send blogs back to Devon, most of Bakau's power having been knocked offline by a virulent thunderstorm the night before, Dr. Chris, Richard, Suzi, Lisa, and I, set off for Gunja Beach in the hopes of being able to confirm the sightings of skinks there. My taxi driver for the journey was a man called Sarah. When I enquired as to Sarah's unusual name, he told me that his father had wanted to give his son a western name, and Sarah had been what he had decided upon. Sarah, himself, didn't mind having a girl's name, and saw it as an advantage in Bakau's competitive taxi industry as tourists never forget him. Conversation ran from the Gambian president's female bodyguards, through to Bob Marley, before we reached Gunja Beach. Walking along the beach, we had to find somewhere to start our search and when, after a while, I found a small hut that looked almost as if it could have been thrown together by the waves, were it not for the rows of charming shell wind chimes with a faded sign saying 'Bar' pointing to it, it seemed clear I had found the Rasta bar Luc had told the others about the day before, so I called them over.

The bar itself was pleasant, and had a genial host, along with a veritable sea of playful kittens in it, which is nothing to complain about. He confirmed the story Luc had told about the skinks, and Richard interviewed him to determine that it was Armitage's skinks he had discovered, and not one of the more common species. So we confirmed Luc's story and by doing this extended the range of Armitage's skinks considerably, which will help conservation efforts of this exceptionally rare and endangered animal.

The next step was to try to catch the skinks on camera, so we headed to the dunes. However, there are some less welcome inhabitants of Gambian beaches, one of which started to follow us without saying a word, along with his dog. This particular vagabond I shall refer to as 'The Creeper'.

`The Creeper` was even more of a hindrance than `Glue` had been on Bungalow Beach, and as we walked into the dunes he and his dog followed us, and our efforts to shake him off by walking though rough grasses and vegetation, as well as climbing over and down dunes simply didn't work; he just followed kept following us in a manner reminiscent of Jason Voorhees in the *Friday the 13th* movies. This more or less dashed our chances of finding skinks ourselves, as the dog would have scared them off. Lisa and Richard asked him what he was doing following us, and told him he was unwelcome as his dog would scare off what we were looking for, and that we were scientists and not tourists, but this didn't seem to deter `The Creeper` all that much, who retorted that he was helping us. `The Creeper` was obviously made of sterner stuff than I, as Lisa works as a bouncer in the north west of England, and I certainly wouldn't talk back to her. After repeated discouragement from all members of the party, and walking further down the beach, The Creeper and his dog disappeared, but by the time this happened the dunes had thinned out and there were no likely skink habitats in this part.

Disheartened, we walked back the way we had come, in the hope that now `The Creeper` and his dog had gone, the skinks might have calmed down enough to show themselves. Our faces fell when we saw `The Creeper` waiting for us (although, this time sans dog).

He wanted to tell us that, if we wanted to find snakes, he had seen a snake by a road two years ago, which was not really of much use to us when we were looking for skinks in the sand dunes, but it seemed `The Creeper` was here to stay. Eventually, Richard managed to get him to sit absolutely still and not make a sound, while I staked out a likely looking habitat. I watched the habitat while shielded from view by a bush downwind of it for over 20 minutes in the fading evening sunlight, but soon it was getting so dark that finding our way back safely would be a problem, so we had to abandon our search without having filmed the skinks.

Day 11, 14th July - `Gambo`, Sacred Trees, Ninki Nanka, Giant Snakes and a Legend put to rest.

This morning we set off bright and early to search for the truth behind two modern-day Gambian legends. The first of these, and potentially the easiest to verify, was that of the Bakau Ninki Nanka hole.

According to local folklore, when a hotel near the coast called the Palma Rima was being built in around 1990, the builders would not commence building work because the area was rumored to be the home of a Ninki Nanka, which would come out of a large hole on a regular basis. No-one was brave enough to hunt the Ninki Nanka, so the hole was filled in while it was away and an established tree planted over the hole, meaning the Ninki Nanka would not be able to use this particular lair, and be forced to move on. According to the legend the plan worked; the Ninki Nanka was never seen in the area again, and work continued on the nearby hotel.

Making conversation with our taxi driver Assan, who works nearby at the Bungalow Beach Hotel (our man in Havana, if you will), I had asked him his opinion on the Palma Rima Ninki Nanka story. Assan was of the opinion that the creature was certainly not Ninki Nanka, but, in fact, a very large python that locals had, perhaps, mistaken for the Ninki Nanka. However, Ninki Nanka or not, the words 'very large python' were certainly enough to keep our interest. After all, the eye witness we found near Dumbletore claimed that the Ninki Nanka he saw was a snake of immense proportions. Walking across the paddy fields behind the Palma Rima, Chris Moiser noticed one tree that was rather different to the others, and halfway between the road and the path, which was where we had been told to look for the tree. We gingerly approached the tree, walking over the small ridges in the fields so we could be sure we weren't trampling a farmer's rice crop, and saw it was a baobab tree. Baobab trees are often considered to be sacred trees, and this tree certainly was, as prayers had been stuffed into knot holes, and even hammered into its trunk using steel bolts. The tree, itself, was clearly over 16 years old, so if this was the tree, it had been well-established before it was transplanted in the hole. To have replanted a baobab over the Ninki Nanka's hole would also have served to calm the fears of the builders, as it would perhaps offer them a sort of spiritual protection from an angry, and newly homeless, monster that, according to rumour, could kill you without even needing to touch you.

From the baobab tree, we walked onwards to investigate the second modern legend of the day: `Gambo`, the mystery creature of Bungalow Beach.

`Gambo` was an animal washed up on Bungalow Beach in 1983. According to Owen Burnham, who was 14 at the time, the strange creature bore a resemblance to a pliosaur and was buried on the beach, 3 feet under the sand and above the tide line. We had some soft drinks in the bar at the Bungalow Beach Hotel while we reviewed Burnham's map of where `Gambo` had been buried, and planned our next move. Our investigation had already been compromised by the fact that the paved beer garden of a nightclub had been built next to the hotel, infringing upon the exact spot, indicated by the map, where `Gambo` had been buried. The investigation was further hampered by the fact that the nightclub was owned by the president's brother; one thing you don't want when digging a 3 foot deep hole in the hope of finding the mortal remains of a cryptid, is the secret police turning up with the army in tow. Try explaining that to a policeman:

"Stop that right now! What are you up to?" the policeman might enquire.

"We're trying to dig up what might possibly be an unknown species of animal," might come the reply, which may have resulted in us being kept in Gambia longer than we'd originally planned.

Thankfully, all we had to contend with was the local beach crowd, that often follow tourists around, pretend to know them and then ask for money. We had had to contend with these people before, and they could be a bit of a pain to say the least. However, this time, as we excavated the sand as near as possible to the cross on Owen's map, we were provided with the perfect opportunity to get our own back on them. Maybe they will talk for years to come about the day the troop of six scientists from the geology department of a British university came to investigate beach erosion in the Gambia by digging a series of 70 cm (3ft) deep holes in the sand, of the short lecture on sand dynamics and erosion Chris Moiser gave them, and of Dr. Chris Clark's 'tricorder'. By the time we had finished the sadly, fruitless search for `Gambo`'s bones the beach boys were offering to help us dig and all wanted to be scientists too.

The story of `Gambo` doesn't end there though. It does seem that it might have been a dolphin after all from what another eyewitness, local shop owner Baka, told us when we asked him about the incident in 1983. Baka was present when `Gambo` was washed up and he said that `Gambo` had been a large dolphin, missing its dorsal fin, which had been washed up on the beach where it had vomited and died. I also acquired a carving of a dolphin in Abuko that looked very similar to Owen Burnham's description of `Gambo`, so it is tempting to assume that the animal was, as so many people have insisted, just a dolphin, albeit a very peculiar one. Sadly, our excavations on Bungalow Beach also showed that the sand is moist from 40 cm (1ft 4ins) down in the area where `Gambo` was buried, so it is likely that the creature's bones rotted away some time ago even before the nightclub was built.

On a lighter note, however, I found out from interviewing staff at the Bungalow Beach Hotel that these days, when large animals wash up on the beach, they are required to call a local vet, who will arrange for them to be taken away and disposed of accordingly. So, presumably, if anything like `Gambo` washes up on a Gambian beach again it will be looked at by a scientist rather than being left to rot under the sand.

Day 12, 15th July - "Psst! Want to see a dragon?"

Earlier in the week, Richard had telephoned in an interview with the BBC World Service about the expedition and this meant that by now tongues had started to wag in the Gambia about the strange westerners searching for dragons on their doorstep.

As a consequence, we got a 'phone call from Abuku National Park this morning, passing on a message from a man, unrelated to the park, who claimed to be able to show us a dragon. The man lived in a village not far from the park called Mandinari. According to his story, there is a dragon that lives in a hole in the mangrove swamps near the village that will come out of the hole if you throw a dog in. The idea of throwing a dog into a hole as food for a possible Ninki Nanka did not best please Richard, so we planned that if the man showed us the hole it would be better to use a chicken instead. We were going to Abuku tomorrow anyway, to see if they could help in tracking down the Englishman who the eyewitness at Kiang West had mentioned, so a trip to Mandinari could easily be added on to our journey. This sounded like quite a fanciful story (if you were a big animal and someone threw a free meal into your hole would you want to come outside of the hole before you finished it?) but if there was a chance the man who had contacted Abuko was genuine, then it would give us a chance to photograph a large draconic animal. Maybe the animal could be the Ninki Nanka, or perhaps a large python, and our appetites whetted, we wondered what tomorrow's excursion could bring.

However, our hopes were dashed as, later in the evening, we received a 'phone call from the man in question. Richard went to answer the call just before dinner and came back after a few minutes looking rather annoyed. It turned out that the man had been incredibly rude and abrupt in his mannerisms and after several insulting comments about the staff at reception being noisy (the staff at the African Village Hotel were nothing short of exemplary in my opinion), the man had tried to negotiate a fee in advance of taking us to this hole. He started with £1000 sterling (about $2000 US) and started working his way up and up and up in £1000 increments, as if it was an auction for the crown jewels, and he was both bidders. Richard told the man he did not have that kind of money and would discuss it with the group. It was clear that the man was some sort of con-artist and, even if there had been a hole he could have taken us to, he would probably have thought up some excuse why the dragon wasn't in residence when we threw in a chicken, and still wanted his exorbitant fee paid. So we agreed that Richard could tell the man that we weren't interested, and that was the end of that.

Day 13, 16th July - The return to Abuko

Despite having turned down the man who claimed that he could show us a dragon the previous day, we decided to visit the mangroves at Mandinari before we went to Abuku National Park anyway, to see if there was anything that we could see for ourselves, and to see if any of the locals had any Ninki Nanka tales - or even better - but we were beginning to think it unlikely that there were any more eyewitnesses that were willing to talk about their experience. There were no eyewitnesses here, but there were large fiddler crabs, that I observed squaring off against each other to defend their burrows, and the supply of females, from other suitors. There was also a local who took a shine to Suzi and held her hand, and professed his love for her. This was a bit fast for Suzi, so, understandably, this was our cue to leave; after all, it is one thing to want a local to give them your watch, but it is, perhaps, a bit too much for them to expect to be donated a member of the expedition team. If they'd have offered us something in return maybe, but sadly thanks to over-generous tourists in the past, and western charities, some people in the Gambia, especially children, expect handouts and something for nothing. [1]

Next, we piled into the minibus, provided by Assan once again, and headed to Abuko, pausing on the way to allow Lisa the chance to realise her ambition of picking a mango from a tree to eat.

At our destination, we enquired at the front desk as to whether they knew anything about Sodokodu, but sadly they didn't and we ventured inside once more.

This time we saw baby crocodiles, and two juvenile monitor lizards in the pool by the pumping station, but we were saddled with the worst guide in the place. I knew I could do a better job of spotting animals, and I also knew how to walk around without scaring animals off with heavy footfalls, so I managed to ditch the others as they went off to a hide, and walked though the jungle on my own.

It was probably one of the best decisions I'd made during the whole expedition, as I saw large monitors crossing the path, was able to watch monkeys at my own leisure and a forest cobra reared up at me on the path. I backed away from the cobra slowly and this showed the cobra I was of no threat, so it retreated back into the bushes - a lesser man would surely have perished. Eventually, I made it to the Ninki Nanka café and animal orphanage, which is located two-thirds of the way through Abuko, and waited for the others to arrive. When they did, the guide asked if he could have my fancy new Casio watch 'as a memento'. Memento of what I asked myself? That he'd been a rubbish guide and I'd ditched him? I declined, and we set off through the jungle to the exit of Abuko, and as I was leading the group, I decided to show off my guiding skills and point out lots of wildlife to the group myself.

Across the road from the exit of Abuko was a charming British pub run by a Rastafarian named Max.

Max has to go on the list with Assan, Musa, Bakari, Lamin (the internet café proprietor) and Sarah, of genuinely helpful and nice people we met in the Gambia, he even was able to converse in Welsh a bit which was a welcome surprise. Max told us that the founder of Abuku, Edward Brewer, when it first opened, had also set up a mirror in the park to discourage the Ninki Nanka, in the same way they had in Papa Jinda's day. He also believed he had his own encounter with the Ninki Nanka in a place called Bintang Bolan, where he nearly drowned and felt like he was being electrocuted. This is not like anything we have heard of before in incidents attributed to the Ninki Nanka, but it is interesting for that very reason. However, some catfish have been known to produce electric shocks when stepped on, so that might be a prosaic explanation for his experience.

---

1. I don't think that Oll means quite what he ppears to here. I must reassure all potential candidates for places on a future CFZ expedition, that we are not in the habit of bartering our female (or male, for that matter), expedition staff, in return for material or intellectual favours. Not unless we really want to that is.

After finishing our soft drinks, Richard and I went to the offices round the back of Abuko where the records are kept, in the hope of being able to access them to find out about Sodokodu and what his western name was, in order to track him down. However, we were greeted at the locked gate by a man who didn't speak English well, and no matter how we phrased our requests to gain entrance, or see someone inside who could help us with our enquiries, thought we were asking for directions to the front entrance of the park.

After another soft drink with Max we went back to the African Village Hotel.

Day 14, 17th July - My final thoughts

Looking at the various witness statements, it seems clear that in the Gambia the name 'Ninki Nanka' doesn't apply to an animal of any one particular appearance; some say it has legs, others that it has wings, or even that it's some sort of serpent of monstrous proportions. There are a few things that have been consistent in nearly all of the sightings reported though, particularly the crest on its head, a large body size, and scales. What is ever-present, and remarkably consistent, is the folklore surrounding the Ninki Nanka; the creature is said to bring death, and is much feared in the Gambia and if you don't die immediately upon sighting it, you can be brought down by a mysterious illness, the only protection from which is said to be spiritual, usually from an Imam. Where traditions of the old animist religions of Gambia have been absorbed into Islam, like at the sacred crocodile pool of Folonko, the Ninki Nanka is sung about when a newly circumcised person takes their first ritual bathe in the river about two weeks after being cut at the crocodile pool, in order to ward off the Ninki Nanka. Practices like this suggest that, perhaps, the Ninki Nanka has been blamed for sudden deaths of people in the Gambia going back as far as when animalist religions were widely practiced in the area, and could even have been the origin of the folklore in the first instance. Of course, more research would need to be undertaken to prove or disprove this theory. On balance, I think that the Ninki Nanka is a mix between vibrant and compelling folklore, and sightings of unfamiliar animals in an area.

`Gambo`, the mystery creature of Bungalow Beach, is likely to have been just a large dolphin that was in a fairly bad way when it was washed up on the beach. Although the bones are almost certainly gone for good, rotted away in wet sand, if not removed during the construction of Destiny's nightclub, we did manage to locate a new eyewitness who identified it as a dolphin, but without a dorsal fin, which is interesting in itself.

All in all, this expedition has been rather fruitful in my opinion, as we have uncovered a lot of new information about the Ninki Nanka, even though it looks as if the mystery of `Gambo` will now, never be conclusively solved either way..

Oll writing up notes

View from the hide at Abuko

A dry watercourse in the jungle

Spotted Hyena (*Crocuta crocuta*)

keeper with hyena pack

Stall at the Nink-Nanka cafe

An Gambian wooden carving of a dolphin

Entering the fishmarket

Drying fish at the market

The market from afar

pufferfish washed up on the beach

Oll Lewis and Chris Clark

Oll tries fishing

female rainbow agama

wreckage caused by Ninki-Nanka?

More of the same wreckage

bridge at the crash site

The ghostly Alahein River

Crossing the Alahein

The empty bush beside the dragon's lair

Momomodu and Bakary

the nightclub's sign in need of repair!

the path behind the Palma Rima

Oll prepares bait

roadside vendor

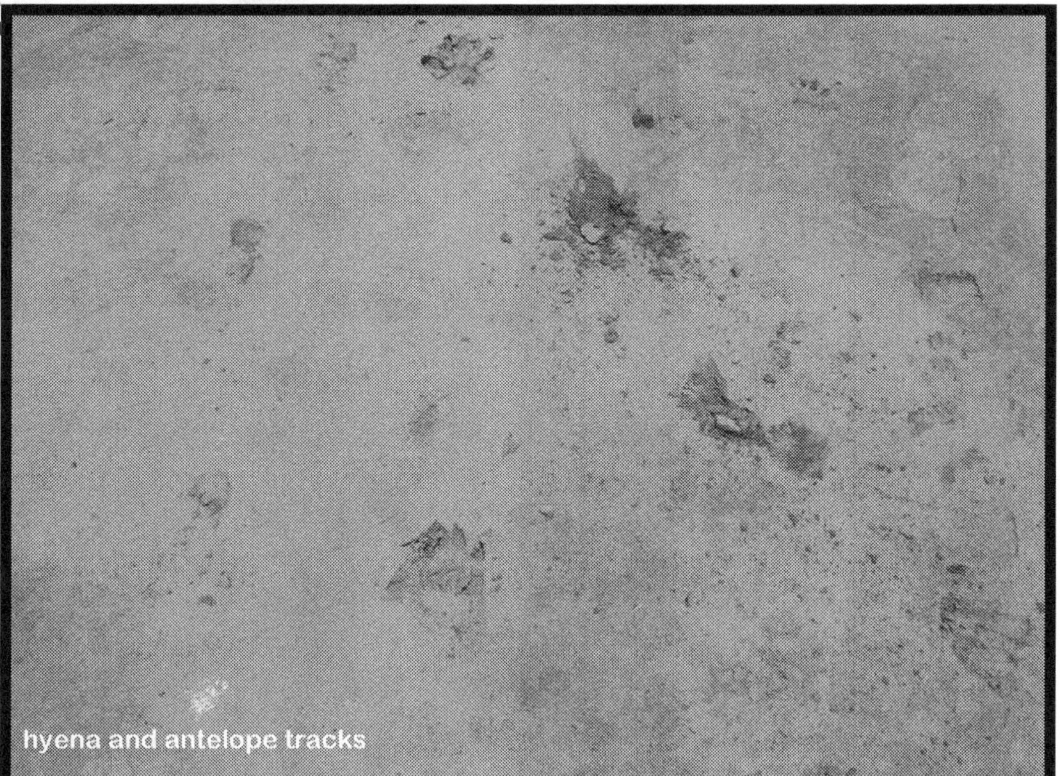

hyena and antelope tracks

Suzi with a fiddler crab

Landscape of Mandinari

Pink backed pelican (*Pelecanus rufescens*)

Suzi Marsh

Suzi writes: "I'm 22 years old and hail from Barnstaple in North Devon. I've always had an interest in Forteana in general, and cryptozoology in particular, and my geographical proximity to CFZ headquarters has prompted me to get more involved in the Centre's activities. After organising a fundraising event for the CFZ, I am delighted to be joining them on this expedition.

I work as a writer for a local advertising agency, and work on various novels and short stories – most with a distinctly 'monstrous' theme – in my spare time."

Four years ago I was on vacation in Scotland with my family. Having already dragged my poor parents, and complaining sister, around the thick hedges that hem Boleskin Manor, former home of Aleister Crowley, they were less than impressed when I proceeded to strike up a conversation with a man who lived in a caravan on the shores of Loch Ness. I had a good chat with this bloke. His extraordinary life and tales of monsters sparked my already Fortean mind down a more specific path of interest. He also warned me off ever entering into a drinking competition with any members of the Centre for Fortean Zoology. Four years later, when I found myself facing up to a spitting cobra and pondering exactly how one such boozy conversation had resulted in this situation, I began to wonder whether it might have been sensible to heed his advice…

* * *

Coming into Banjul airport over strips of grubby mangrove is an accurate introduction to how the wild jostles with the urban in the Gambia. As soon as we stepped off the plane, a butterfly zipped in front of our faces before alighting on the window of a bus the wrong side of dodgy-looking. At our hotel we were greeted by smiling faces; Nike trainers stuck out from under traditional dress, agamas and skinks scuttled happily through air con units and across crazy paving. The first thing I noticed was how happily these things that should jar actually sat together; how the modern seemed perfectly at ease with the old. Which is handy, I suppose, when you're entering an area that's spiralling headfirst into modernisation, intent on getting the facts about an ages-old monster that stalks the rapidly decreasing mangroves.

After getting our bearings and recovering from a long journey, we set out on Wednesday 5th July on the first proper day of our expedition. We took a stroll down to Bungalow Beach, site of the carcass burial reported by amateur naturalist Owen Burnham, who was 14 at the time. I have no doubt that the story of this beast will be described elsewhere in this book more expertly than an amateur such as myself could manage, so I will spare the back story. I will only mention my own perception of the legend; that it is one of those wonderful stories with the power and intrigue to survive despite all the evidence backed up against it, and that I strongly hoped we could bring it to its conclusion one way or another.

We noticed a problem as soon as we got to the beach. What had once been a remote stretch of undisturbed beach, no doubt fertile ground for any number of mysteries and curious beasts, was now a commercial strip, the crowning glory of which was *Destiny's* nightclub, rising obstinately from the very spot we had come to explore. We did not have the map drawn by Burnham on us at this stage, so couldn't be sure of the carcass' exact location, but the building, and the fact that it was owned by the President's brother, certainly seemed to cast a shadow, both literal and otherwise, upon the proceedings.

Travelling on up the beach we came to Kotu Tourist Market where, as previously reported in *Fortean Times*, local silversmiths create jewellery in the shape of Ninki Nanka. At first glance the designs seem similar to a stylised Western dragon, but there are minor differences; most significantly the addition of a crest on the creature's head. One of the locals selling this jewellery was Baka Samba, a fountain of useful information. Seeing this man talk was the first time I had experienced the discussion of something so fanciful to Western ears in such earnest, fearful tones. He spoke of Ninki Nanka, and of how his uncle had encountered the beast out hunting. He described it as very big and terrible, with four legs, a long tail and fire in its mouth. Baka claimed that anyone who saw the Ninki Nanka would suffer ill health, and pass away within five years. Later, in discussion with the rest of the group I began to discover the fascinating process of applying logic to the most outlandish of claims; that 'fire in the mouth' could well mean a mouth or tongue that is red or orange in colour; that the 'curse of death' brought on by seeing Ninki Nanka could be the result of a disease, picked up in the deep swamps where the beast is said to lurk and lying dormant only to later re-surface, with tragic results. I began to realise that this was what Fortean Zoology was all about; not believing the unbelievable whatever it may be, or scrawling 'here be dragons' with abandon on any given map, but looking for logical, scientific answers where mainstream research has turned its head.

* * *

The next day we headed for Abuko Nature Reserve, an area of gallery forest and Guinea savannah which plays host to a staggering variety of wildlife including many species of monkey, Nile and dwarf crocodiles, Nile monitors, numerous skinks and geckos, a variety of both poisonous and non-poisonous snakes and more birds than you could ever attempt to count or catalogue.

Abuko was the setting for one of the most famous Ninki Nanka sightings, reported as it was in a *Lonely Planet* guide to the area. The creature was said to have destroyed the reserve's pumping station, which we later saw lying in a state of photogenic disrepair. We gathered a lot of information here. Our guide, Musa Jatta, a genuine Dr Doolittle, with the ability to perfectly mimic the calls of various birds, told us of an accident that occurred on a road upriver where a lorry crashed trying to avoid a Ninki Nanka. He described it as a large python with legs that could swallow a cow whole. He said it had bat-like wings and could breathe fire. Hassan, the grandson of Papa Jinda, who witnessed the incident at Abuko, told of his grandfather's experiences; of how the staff at Abuko had panicked at the appearance of Ninki Nanka and requested a mirror, the only way to kill the beast, and of how Papa Jinda encountered the creature for a second time which was to prove fatal. Hassan said his grandfather became very ill, complaining of pains in his legs and waist, before his hair fell out and he died two weeks later. He described the animal as having a 'tiara' of fire on its head. We also met an elderly acquaintance of the staff at Abuko who claimed to possess scales from a Ninki Nanka, which he claimed glowed red at night. We also had a report from the security guard at our hotel, who claimed Ninki Nanka had a crest bearing Islamic scripture upon its head.

Throughout this day and the last, I gradually began to notice the change that seems to take hold of people when they talk about Ninki Nanka. The cynic would say this is nothing more than a cleverly honed act, that secret sniggers are being indulged in behind closed doors at the expense of the crazy English people who are looking for dragons. But, in my opinion, a change does take place in people when they talk about Ninki Nanka. Maybe it is nerves that their outlandish tales will be dismissed, genuine fear born from generations of folklore or just plain theatrics, but a certain animation does seem to creep into their manner; not a false tourist-pleasing grin but a true intensity coupled with uncertainty, a trait glaringly omitted from the rest of their speech. Eyes still twinkle and laughter still flows, but a genuine fear seems to lie beneath.

On a lighter note, I saw many animals I had never seen in the wild before, including crocodiles, monkeys and a rather terrifying spitting cobra. My fear of creepy crawlies, present since childhood, completely vanished in the face of so many new and fascinating things. As anyone who knows me would tell you, for me to happily gawp at various 'pedes' of unknown numeration is nothing to be sniffed at...

* * *

Kiang West National Park is, without a doubt, the most beautiful place I have ever seen. Some of the more experienced members of our party, and indeed the *Guardian*'s online blog, may scoff at the way its wild appeal was slightly marred by the presence of a rather less than wild picnic area, but for someone who has never travelled far, an open expanse of savannah broken only by the prints of hyenas and baboons is about as exciting as it gets.

On the way to Kiang West we stopped at the site of the crashed lorry. A rather forlorn wreck, it lay on the roadside in a suspiciously advanced state of decay while, some 200 feet up the road, the 'bridge' built over the furrow created by the Ninki Nanka's crossing bore a strong resemblance to a drainage system. My instinct told me, however, not to immediately dismiss the story. I have no idea how far a lorry could roll in muddy, wet conditions, nor how many times the wreck would be rolled by locals in search of anything they could use or sell. This is something I wish to research further. It is also worth noting that along the way we passed scores of crashed vehicles, many of which, were I creating a story of a crash caused by a mighty creature, I would have chosen as better fitting the bill.

At Kiang West our guide Bakary took us as far as he was willing to go and pointed us in the direction of the mangroves. We followed the 'track', which in places was little more than a place where a few tree branches had been pushed back, to the site of village abandoned, according to Bakary, in fear of Ninki Nanka. Sadly, we didn't see anything, but certainly felt an ominous atmosphere and a stillness that felt odd even in the mid-day heat.

The most important incident this day was our meeting with Momodo, a man from Bakary's village who claimed to have seen Ninki Nanka himself. He pointed out holes in the ground from which he had seen the creature emerge, and took us to a point along the water's edge where he said he had seen Ninki Nanka. Unusually this, our only first-hand witness account, was the most outlandish; Momodo described the creature as 50 metres long and 1 metre wide with the face of a horse and the legless body of a snake. When we asked him about its teeth and tongue, he said its mouth was closed so he could not see, which in itself felt unique. Surely, if someone were out to tell us what they thought we wanted to hear, they'd have fit their description to the Ninki Nanka's popular appearance? Momomodu also said Ninki Nanka had mirrored scales, a many-hued, predominantly green skin and a feather-like crest which hung down over its face. Later, on being shown pictures of a Chinese dragon, a Komodo dragon, a brontosaurus and a Nile monitor, and asked which one best described what he had seen, Momomodu picked out the Chinese dragon, saying its face was that of the Ninki Nanka. He also said that seeing the Ninki Nanka made him very ill, but that he had been cured by a marabout who gave him a herbal potion.

Momomodu also gave us another lead which we are currently following up on; the report of an Englishman, working for the Gambian wildlife authorities, who had camped out in these mangroves for five days looking for Ninki Nanka and had, apparently, been successful in his quest. I feel this is a very important lead for us to focus on; that a report from someone who is removed from the lifelong folktales of the beast which are omnipresent in Gambian society would give a new level of credence to Ninki Nanka's existence. The stories we had heard up to this point, and would hear throughout the trip, were always irreversibly intertwined with belief. Children grow up with Ninki Nanka, the stories so ingrained that any large snake or other reptile could appear to someone in the right mindset as the monster. If we can trace someone without this background, a wildlife expert, who has seen it, we can begin to move Ninki Nanka from folklore to something more tangible.

* * *

Banjul was the biggest contrast to Kiang West we could have found within the Gambia. And yet, even here, amidst bustling markets, hollering traders and all the trappings of modernisation, Ninki Nanka still lurks in the public conscience, albeit in a different form. Here we were hunting for something slightly more mundane; a tape recording of a 1980s hit song by a band called *Toure Kunde*, entitles *Ninki Nanka*. In this search we were successful; not the most exciting of revelations, but a nice little side note. Although genuinely believed to be a source of great fear and certain death, Ninki Nanka is still thought to be suitable fodder for a cheery Gambian pop song.

* * *

At Kartong. we took a trip in two dug-out canoes to take a closer look at the mangroves, favoured habitat of the Ninki Nanka. The water was high, and in places thick roots clogged the water below the surface. The mangroves are dense and foreboding; it is easy to see how legends of a hidden creature have grown up to surround them.

Our 'captain' told us that Ninki Nanka can "metamorphise"; that it can appear very large one second and merely knee high the next. Interesting as this was, it drags the creature firmly back into the land of folklore or paranormal origin, rather than the animal of flesh and blood its alleged sighting by the wildlife expert at Kiang West may suggest. The contradiction surrounding different reports of Ninki Nanka is ever present; by this point I was beginning to suspect the reason the creature is so elusive is because it's in the midst of an identity crisis. The poor thing doesn't even know if it's got legs or not.

At the Gambian Reptile Park we found success in another aspect of our expedition; the hunt for specimens of Armitage's skink. Although Luc Paziand, the charming Frenchman who owned the park, sadly did not have a specimen, up until recently he had kept a pair given to him by the owners of a bar on nearby Gunjur Beach. We also chatted to Luc about Ninki Nanka and he put forward a theory that the animal could be a large snake with a growth or leeches attached to the crown of its head, giving the appearance of a growth.

At Bungalow Beach, stopping to talk to an acquaintance of our driver Assan named Lami, we received information that pointed towards Guinea as being the possible site of a more prolific Ninki Nanka population. We had had hints towards this before; comments here and there that in Guinea it was more common to see a Ninki Nanka. The interesting aspect of this man's tale was that the fear Gambians have towards Ninki Nanka is not universal; in Guinea, people don't seem to fear it or believe it will cause death. Indeed, Lami claimed his grandfather 'kept' a Ninki Nanka as some kind of extremely exotic pet; that he could summon it to him at will. Lami said the creature looked like a snake, but with a round, 'ball-like' head. Sadly, Guinea is too far away from the Gambia to have been a plausible detour on this trip, but we have had an invite from Lami that if we decide to investigate Guinea, an option which looks promising as this area of Africa is much more unexplored, we can arrange a meeting with his grandfather and, hopefully, experience this summoning first-hand.

* * *

The next day, we decided to take an early evening trip to Gunjur, a long strip of dune-backed beach which Luc Paziand had told us was a habitat of Armitage's skink. With surprising ease we located the beach bar whose owners had donated two specimens to the Gambian Reptile Park, and were able to confirm that they had, indeed, captured the creatures. They described them as brown, about 10 inches long and as thick as a finger; a description which matches that of Armitage's skink. Sadly, this was all we were able to discover on this expedition as the attention of various 'bumsters' and their accompanying shouts/animals made sure that any animals that might have been lurking were long gone before we could get anywhere near.

* * *

Legend has it that when the Palma Rima Hotel at Bungalow Beach was being built, a Ninki Nanka was discovered living in a hole nearby. Fearful of the beast, workers waited for it to leave its hole before planting a large tree over the entrance, preventing the Ninki Nanka's return and leaving the coast clear for the construction of the hotel to continue. Our driver Assan, who works at the nearby Bungalow Beach Hotel, was certain that the story was a semi-truth; adamant that, although a creature did halt the building work, it was a giant python, not a Ninki Nanka. This still interested us; if we could prove the location as a habitat for a giant python it would give suggest that locals, although familiar with snakes on a fairly regular basis, are still capable of confusing one with a Ninki Nanka...

Lying exactly in the location Chris Moiser had been given – halfway between the road and the cycle track beside the Palma Rima – was a baobab tree of immense proportions. We gingerly picked our way towards it, through fields that have been known to conceal the odd poisonous snake, and were immediately struck not only by its size but by the heavy atmosphere that seemed to surround it. This tree was clearly very old, which immediately shed doubt on its identity as one which had been moved in 1990, when the Palma Rima was built. It would have taken a lot of people, with a lot of dedication and possibly even mechanical equipment to have been able to move this tree; this, however, is not completely implausible. If locals were fearful enough of the creature they may well have chosen this tree to protect them – all baobabs are seen as sacred by some in the Gambia, but this one seemed overtly so. Geometric carvings marred its smooth surface, in some places barely distinguishable from the natural lines of the tree itself, and the bark was bursting with scraps of paper, rolled up and crammed into cracks or bolted onto the tree, which we assumed were prayers, though of course were not rude enough to take a closer look. This led us to a 'chicken and the egg' situation; had the tree become this sacred as a result of its alleged banishment of a rogue Ninki Nanka? Or was it chosen as a suitable tree to block the hole *because* of its revered status? If the latter, would the authorities allow such a sacred landmark to be uprooted and blasphemed in such a way? A cursory glance at the surrounding area, with its carpet of coke cans and broken glass, suggested that yes, they probably would.

Interestingly, while researching sacred baobab trees in the Gambia back home on the internet, I came across a report from a BBC correspondent in Makasutu, near Mandinari, of a dragon that was said to live under a baobab tree in the park. (Source: http://news.bbc.co.uk)

From here we trekked on to Bungalow Beach to complete the investigation of 'Gambo' which had proved so difficult before. We had reasoned that any digging would be best done towards the end of the trip; should we attract the attentions of authorities we would hopefully be well on our way before accusations could be made...

Sadly, our suspicions that anything buried at Bungalow Beach had long been lost to both natural and man-made interference were confirmed. Digging beside the wall of Destiny's nightclub, at the spot indicated by Burnham's map, we found that much of the area was reconstituted sand, presumably thrown up and reburied when work on the nightclub was taking place. Anything that was there to be found would have been discovered during the construction of the building's impressive foundations, and presumably disposed of. We already know that burials on this beach are a regular occurrence, so, to any of the (independent and, sadly, untraceable) contractors working on Destiny's, a large carcass would have been nothing to write home about.

It is worth noting, before anyone launches into a diatribe on how this admittedly rather ugly, commercial building has ruined society's chances of solving a famous cryptozoological mystery – as I admit I was tempted to do – that, while digging a series of 3ft deep holes, we observed the water level to be a mere 40cm down; an ecological circumstance likely to have rotted away any remains before the nightclub got around to interfering. I feel that, with the possible exception of being able to track down more witnesses to the burial itself, the mystery of `Gambo` has been taken about as far as it can go. Although we have gathered enough evidence to hint at the creature's identity – eyewitness reports that it was a dolphin missing its dorsal fin – Bungalow Beach seems determined that its secrets will remain such, and it's likely that, in true Fortean style, we will never be sure of the truth.

* * *

A man from Mandinari, an area of thick mangroves near Abuko, telephoned us to say he could show us a hole where a Ninki Nanka lived. When he demanded a fee so outlandish I almost thought he must be for real, we declined his offer but decided to take a trip into Mandinari ourselves. The main reason for this was to photograph and document these areas, to show that there are still unexplored mangroves plenty large enough to conceal a secret or two. This area is also very close to Makasutu 'Culture Forest', now a tourist trap but in the past a mystical place more synonymous with its name, which means holy or sacred forest in Mandinka. Legend has it that a Ninki Nanka once lived here, and Chris Moiser speculated that the ever-growing tourist trade could have forced it to abandon its home for the quieter mangroves of Mandinari.

On arrival, we saw a thick mass of intertwined roots and leaves, dangerously concealing muddy waters from unsuspecting feet. A rickety palm wood walkway stretched into the distance; standing rather unstably on it I was rewarded with a good view of the landscape that surrounded me – thick vegetation as far as the eye could see, broken only by the occasional bend of snaky, brown river. A few jetties such as these were the only signs of civilisation, and these were barely standing under the weight of the local children who, rather bravely, used it as a platform for swimming and diving. The purpose of the brief trip had been fulfilled – we were now without a doubt that a habitat sizeable enough to conceal Ninki Nanka certainly exists, and within striking distance of one of its more documented homes.

We went to Abuko to take a farewell trip and to attempt to track down the Englishman Momodo had told us about at Kiang West. We had hoped that the employees at the reserve might recognise the man, but sadly we could not find anyone who did. We hope to make further attempts to trace the man, who went by the Gambian name Sudokudo or similar, on our return to England.

Eleven baby crocodiles and a troop of red colobus monkeys later, we came out on the other side of Abuko, and into the charming bar of an equally charming Rastafarian by the name of Max. He told us of an experience he had as a child at a place called Bintang Bolon, similar to what a westerner may well deem a 'near death experience'. He said he was swimming in the water when he saw 'electricity' bouncing on the surface, before the water turned into a whirlpool and dragged him under. He said he could see his own body before he was revived, and to this day refuses to swim in the river. He equates this experience with the presence of a Ninki Nanka in the area, and says that this is a place where we "might find what we're looking for". Max also told us that Mr Edward Brewer, the founder of Abuko, had, after the incident at the pumping room, erected a giant mirror, with the apparent aim of killing the Ninki Nanka should it decide to return.

*   *   *

In conclusion, I do feel that the expedition has been a success, despite not returning with the head of `Gambo` or photographic evidence of the Ninki Nanka. We have, however, amassed large amounts of data relating to the position of Ninki Nanka in the Gambian conscience, including both first hand accounts and examples of stories passed down through generations, and have also come away with a substantial amount of information on modern sightings, which were few and far between before this trip. We have also taken things as far as we feel they can be taken with the mystery of Bungalow Beach. As seems to be often the case with such things, in finding answers to old questions we have come away with many new ones. We have some exciting leads on Ninki Nanka that we can follow up on our return to the UK, and most of us have developed our own avenues of thought as to what the creature could be. Personally, I believe it could be a case of occasional sightings of exceptionally large snakes, possibly with a deformity or other natural anomaly causing a crest-like formation on their heads, combining with local folklore of a dragon-like being; although, in true Fortean style, I am open to speculation and certainly don't see this as a given conclusion. One thing that is certain, however, is that in the minds of the Gambian people the Ninki Nanka is as real as anything else, and still, in the 21st century, a cause of great fear and suspicion.

*   *   *

And so concludes my first adventure with the CFZ. I have found it to be a very enlightening experience, my previous trips abroad having consisted mostly of lying very bored on a beach while my friends sunned themselves. This trip really opened my eyes to hands-on research; although it was relatively exploration-light in comparison to other CFZ expeditions, it made me realise that not everyone with an interest in cryptozoology, or indeed Forteana as a whole, sits at a computer analysing photos and arguing the minutest detail. It is comforting to know that people are getting out of their swivel chairs and into the field; a feeling it seems is shared by the many encouraging posts I have read on the internet, both before and after our trip. The expedition seems to have inspired people, and I feel that this, along with the high level of media interest which has brought cryptozoology and the CFZ to a whole new generation, are the most significant successes of this trip.

Richard at Gatwick airport

The sight of an alleged encounter with Ninki Nanka

the road the dragon was supposed to have crossed

The entrance to Kiang West National Park

Woodland in Kiang West

The dry bed of a seasonal lake

The River Gambia

Abandoned huts

Jungle in Kiang West

A troop of common baboons at Kiang West

Suzi writing up her notes

Suzi examins the trench supposedly left by Ninki Nanka

Eerie derelict huts

Suzi at the Gambian Reptile Park

Mangroves at Kiang West

Displaying agama

Suzi's toilet being fixed!

Great Egret (*Ardea alba*),

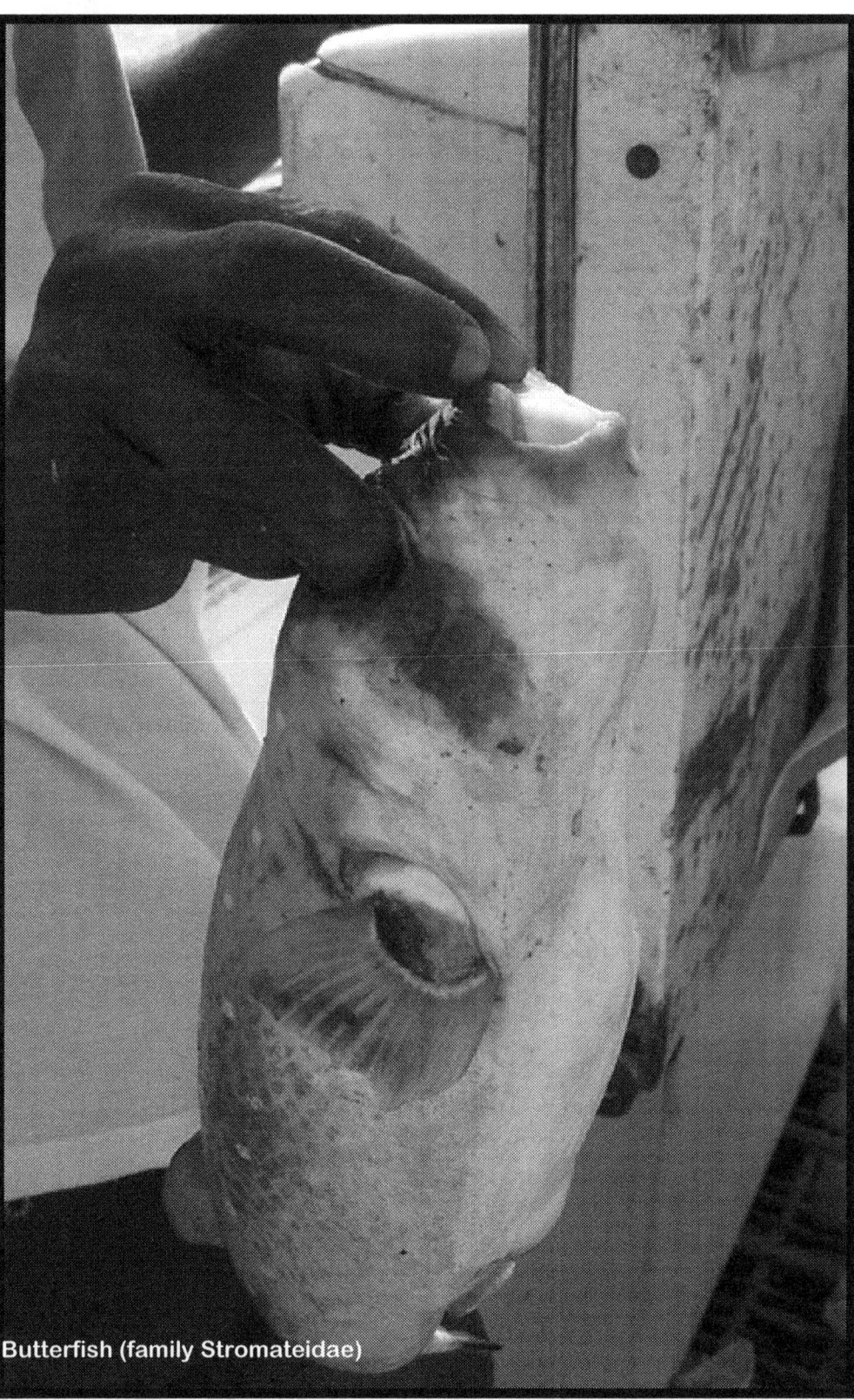
Butterfish (family Stromateidae)

Chris Clark

Dr Chris Clark, Born 9 months after VE day, I can claim to be among the first of the Baby Boomers. While working for a Ph.D. in astrophysics I was involved in studies of the first lunar samples brought back by Apollo 11. I have researched in solar astronomy and atomic physics before settling down to write engineering software. A childhood interest in cryptozoology led me to take expeditions to Sumatra in 2003 and 2004 in pursuit of orang pendek, and in 2005 to Mongolia. In 2003 I also joined the first tour of Afghanistan for 25 years; after that experience, all other destinations should be easy!

July 5th

Walked out to Bungalow beach in the morning to check on the site of the 'Gambo' burial. It now appears to be mostly covered by a night club belonging to the President's brother. Looked at the local market and spoke to a jeweller called Baka Samba who claimed his uncle had seen Ninki Nanka. He said it was 'dangerous' and horrible', and that it had a red mouth or tongue. He said he had been there in the 80's when an animal was washed up on the beach and identified it as a dolphin, but did not see any dorsal fin.

His neighbour knew of a lake in the Casamance area of Senegal. A large lake, of unspecified size but presumably several kms at least, it is full of vegetation. It has disturbances of the water surface and lights seen at night. Also tracks can be seen in the area around the lake – no footprints. The lake is full of rotting vegetation and smells bad – possibly methane. The locals are afraid of the area and leave when the lights are seen. We discussed the possibility of going there. It would involve a long drive upriver, crossing the river and returning back west before crossing into Senegal. For some reason the car ferry was out. I think we should only bother with this if no better expedition turns up. Presumably we would have to stay by the lake overnight.

July 6th

Trip to Abuko Nature Reserve. Saw various species of bird from an observation platform by a pool. The walk was interrupted at one point by a young spitting cobra which caused the people in front to move smartly backward. Afterwards met a man who claimed to have a dragon scale. Richard bought it for 300D (£6). Looked like a piece of mica, but the texture was more like plastic. We found a witness with a story about his grandfather who saw Ninki Nanka twice in the 1940's. The first time it had broken a large water main and he discovered the hole. After the second time his legs and stomach swelled and his hair fell out; the suggestion of radiation poisoning caused Richard to mention Godzilla. We also encountered a man called Bakary Jarju who claimed that N-N had appeared at his village – Matele? [seems to be Batelling on the map]. Since this is closer than Casamance it might be worth going.

July 7th

Right shoulder froze in the night. Woke up and could not sleep – took 2 aspirins. Got up in time for breakfast and took 2 Ibuprofen – shoulder now not painful. In the morning went for a swim in the sea - very warm, sun shone, good waves, rocks on sea bed so needed shoes. Exercise probably good for shoulder. In afternoon walked from hotel to fish market with Richard and Lisa. Wide variety of fish particularly large barracuda. Fish frozen, smoked, salted etc, very strong smells everywhere. Picked up tout who insisted on accompanying us. Gave him 200D for 'bag of cement' (original at least) when police appeared with hotel employee and drove him off. Hotel guy proved to be worse parasite than tout, collecting 300D from each of us as 'wedding gift' on grounds that he was recently married. Impossible to have any respect for people here: I always expect any conversation to be brought round to the subject of money sooner or later.

July 8th

Started at about 6:30 to take a long journey in an open long-wheelbase Land Rover with Richard, Suzi and Oll and Bakary Jarju to check on a location where the Ninki Nanka is believed to exist. On the way we stopped at location [13° 20.066'N 15° 49.885'W] where N-N is supposed to have crossed the road leaving a furrow in the ground and causing a lorry to crash. We found an abandoned container about 150 yd from the site of a concrete drainage culvert. The culvert is said to have been built over the furrow caused by N-N: however, the crash was only 2-3 years ago and the culvert looked older. Also much too far from apparent crash site.

We reached a point close to the Gambia river [13° 26.765'N 15° 50.698'W] where N-N was supposed to live. Oll struggled on the short but very hot walk across a piece of open savannah towards the mangrove trees on the river bank. The local guide hid behind a tree from fear of seeing N-N: Bakary Jarju went a little further to give us directions and then also hid. We followed the thin fringe of mangrove trees around the edge of the river for about ¼ mile before reaching a picnic spot where we ate the packed lunch the hotel had provided, and then came back. (Apparently there was going to be some tourist facility there but it was abandoned before being completed.) It is impossible to imagine a 30' reptile in such a spot: the whole thing felt ridiculous, like a trip to Box Hill with the children. I don't envy whoever has to write this up as a cutting-edge expedition. Certainly it is a come-down from the Lost Valley in Sumatra, or the Gobi Desert. Next year Tajikistan or Siberia.

We went back and stopped at the village of Batelling, which I was surprised to see appears on the GPS map. A local man claimed to have seen N-N and to have a scale. He took us through fields and pointed out holes of various sizes which he said were the home of N-N. Some were so small that I wondered if Ninki-Nanka could also be the name of a ground-dwelling rodent. Apparently the holes had shrunk with time. Finally took us to a spot [13° 23.294'N 15° 51.529'W] which was a creek of the R. Gambia to show us where he had seen N-N on the opposite side. He described it as 40-50m long (this was definitely meters not feet), no legs or wings, body round like a snake, moved slowly, large crest on head, scales various colours but reflecting light like the sequins on Suzi's clothes, especially on the crest. Since it did not open its mouth he saw no teeth or possible red tongue. Gave him 75D. He offered us a look at his scale but we have seen enough plastic so told him that we needed to leave at once to get back on proper roads by sunset.

Returned about 8:30. Driver got 5500, guide 600D plus ride back to his village. Straight line distance was about 90km but much further by road: 140-150km.

July 12th

Off at 9:00 by taxi to the far south. I called the American Express office before we left Bakau, but was told that they could not supply any money. Went on the Allahein river in 2 dug-out canoes (Oll in one and me, Suzi, Richard & Lisa in the other). Carried us around the mangrove swamps and landed in Senegal for a few minutes (the river is the frontier here), though very little wild life. We called in at Follonko in the Kartong district to see the sacred crocodile pool. We had to remove our shoes like a mosque but nothing else Moslem. Too hot to see crocs except for a single very small one. Legend of a white croc: if you see this you will soon become a big man. The pool is used by women before circumcision ceremonies, or when wanting children.

Then on to the Gambia reptile park. Held a Ball (Royal) python, also African egg-eater and shovel-headed snake. They had spitting cobras, puff adders etc, also a pelican and a baboon. Owner – Luke Paziand (French) lives there with wife (Gambian) and 4 children including small boy who clambered over all of us. He is apparently thinking of moving to Guinea because of the Gambian authorities.

Finally to Bungalow Beach to speak to a man called Lami. His grandfather in Guinea-Bissau had seen it and described it as like a snake but with a round head. Also said that he (grandfather) could summon it – like an African Doc Shiels.

July 15th

One thousand hits/day on the blog. Great stuff, but I wonder if a bigger, and more drawn-out expedition, possibly looking for a less media-friendly creature, would generate as much interest. I can imagine us next year operating at 11,000' by a glacier in Central Asia, looking for creatures that almost certainly exist, and entirely failing to get half as much attention.

Everyone is agreed that the biggest problem operating here is touts. Everywhere you go in the city someone will sidle up to you and offer to guide you somewhere, sell you something or take you to someplace you don't want to go: no matter what you say or do they will not GO AWAY. We were looking for a lizard yesterday, and someone offered the services of himself and his dog We had another go at digging up the skeleton of the sea creature yesterday (the one next to the night club) and pretended to be scientists doing a beach erosion survey. We had a great time: I pretended to 'scan' the beach with my GPS while pressing enough keys to make it beep, at the same time talking to my digital watch, being sure to press its buttons, and address it as 'Tricorder'. All this time we were digging with kiddies plastic spades. Gambia, which used to be a fairly decent place is becoming more a bog-standard one-party state: there is a treason trial going on as a result of an attempted revolution a few weeks ago, and there was a good deal of military activity this morning. I'm surprised, in view of some of our activities during our visit, that we weren't suspect too.

Suzi at the railway station in the wee small hours

Oll at Gatwick airport

Chris Clark in the jungle

bucket traps in Abuko

Vervet monkey at Abuko

Laughing or spotted hyenas

Mangroves of the Gambia from the air

Mad dogs and Englishmen really *do* go out in the mid day sun

Chris Moiser speaks with a waiter from *Destiny's*

The team on the beach

The beach of black volcanic sand

Smoking catfish at the fish market Baku

The poorer fishermen sort out their catch

Gambian fishwives

Oll in a dugout canoe

Ninki-Nanka's home?

The road to Kiang West

Exploring the sandunes of Gunjur

Goliath heron (*Ardea goliath*)

Bosc Monitor (*Varanus exanthematicus)*

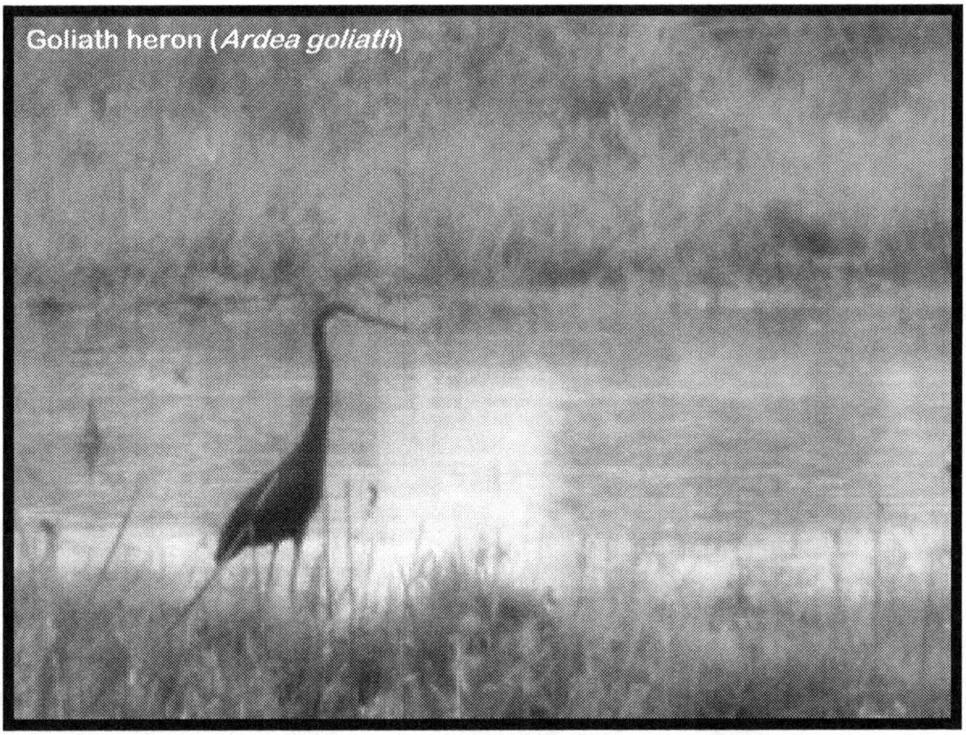

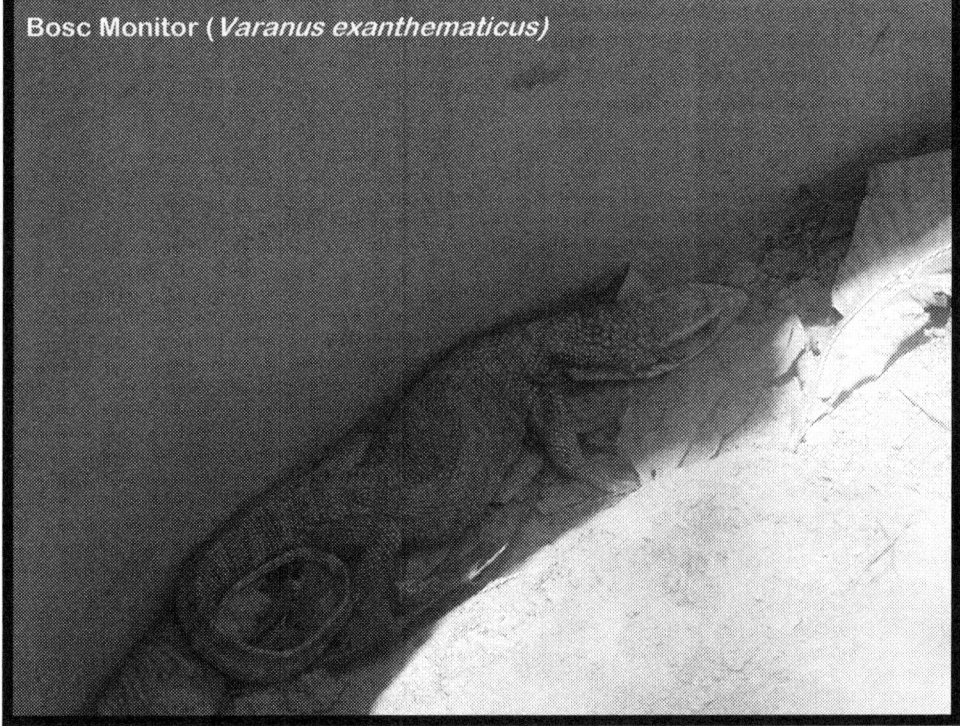

The team's last day in Gambia - on the veranda

APPENDIX ONE

CRESTED SERPENTS WORLDWIDE

Richard Freeman

'The serpent - subtlest beast of the field'
Milton, *Paradise Lost*

The serpent bearing a crest, fin, or comb, akin to that of a fowl is a motif that occurs again and again in the world's folklore and in modern sighting of strange creatures.

Perhaps the best known is the basilisk. Of all dragon-kind, the basilisk is probably the most arcane. Its genesis involved a series of events so unlikely that, (luckily for man), they conspired only rarely to create one of these baleful monsters. It was believed that occasionally - in old age - a rooster could lay an egg! If such an uncommon egg were to be incubated by a snake or toad, then a basilisk would hatch out - to bring death into the world!

The basilisk was one of the smallest of the dragon-tribe but one of the most lethal! Its death-dealing powers came not from fiery-breath or tooth and claw but from its withering-glare. Any creature that caught the eyes of the basilisk would fall dead from the uncanny-power of its vision. There was but one exception to this - one animal that could withstand this 'look of death'. That animal was the weasel! It was believed that God never created a bane, without creating some cure for it, (like the stinging-nettle and the dock-leaf). Ergo, someone who knew its weaknesses could tackle even the basilisk. The monster's own gaze was as lethal to itself as to any other creature. Hence, its own reflection would kill it stone-dead! Equally - for some cryptic reason -

---

Marcus Atilius Regulus (died c. 250 BC), a general and consul (for the second time) in the ninth year of the First Punic War (256 BC). Regulus defeated the Salentini and captured Brundisium (now Brindisi) during his first term as consul in 267 BC.

He was one of the commanders in the Roman naval expedition that shattered the Carthaginian fleet at Cape Ecnomus, and landed an army on Carthaginian territory. The invaders were so successful that the other consul, Lucius Manlius Vulso Longus, was recalled to Rome, leaving Regulus behind to finish the war.

After a severe defeat at Adys near Carthage, the Carthaginians were inclined towards peace, but the terms proposed by Regulus were so harsh that they resolved to continue the war. The Cathaginians replaced the outmatched general Hamilcar (Drepanum) with new leadership and in 255 BC, Regulus was completely defeated at the Battle of Tunis. He was taken prisoner by the Spartan mercenary general Xanthippus along with 500 of his men.

There is no further trustworthy information about him. According to tradition, he remained in captivity until 250 BC, when after the defeat of the Carthaginians at the Battle of Panormus he was sent to Rome on parole to negotiate a peace or exchange of prisoners. On his arrival he strongly urged the Roman Senate to refuse both proposals, and honored his parole by returning to Carthage where he was tortured to death (Horace, *Odes*, iii. 5). This story made Regulus an example of heroic endurance to later Romans; most historians, however, regard this account as insufficiently attested, as Polybius does not mention it. The tale may have been invented by the annalists to excuse cruel treatment of the Carthaginian prisoners by the Romans.

Of the COCKATRICE.

the sound of a cock crowing at dawn would also kill the basilisk.

These monsters came in a variety of shapes. They first appeared in the bestiaries of the Dark Ages. It was described as a tiny serpent, about a foot in length, bearing a crown or crest upon its head to denote it as The King of Serpents. The deserts of North Africa and the Middle East were reputed to have been created by basilisks, whose glare was so terrible that all vegetation withered under it, and even solid rocks were split and sundered into sand.

Later reports of basilisks came in from Europe. As the centuries passed, the basilisk's form changed. Sometimes it was portrayed as a lizard with a rooster's head, or as a large lizard with six legs and a crown upon its head. The commonest form that these later basilisks took was that of a huge-rooster with the tail of a serpent or a lizard. Sometimes these beasts sported horns or antlers. In this form, they were known as the cockatrice.

In Britain the town of Saffron Walden in Essex was supposed to have been terrorised by a basilisk. It was only a foot long, black and gold in colour and bore a crown-like marking on the head. It was said to be able to strike living things dead with a glance, even from a distance. It held the town under siege until a knight, who wore armour covered in crystal glass, killed it by showing it its own reflection.

In Wherwell, Hampshire, a cockatrice was said to have been created when a toad hatched a duck's egg in the ruins of an old priory. The resulting monster terrorised the area until an enterprising youth lowered a mirror into its lair, destroying it.

A factor in the legend of the basilisk seems to lie with the black mamba (*Dendroaspis polyepis*) - a large and highly-venomous snake of sub-Saharan Africa. This snake can reach four metres (fourteen feet) in length, and is renowned for its aggressiveness. When in a warning display, it can rear up to the height of a man, and, unlike most other snakes, it will actively pursue and attack anything that it perceives as a threat. Most snakes strike only once. The black mamba strikes repeatedly, and has potent neurotoxin (nerve-paralysing) venom. It has been recorded, on many occasions, that black mambas shedding their skin will sometimes retain a flap of old dead skin upon their head. This strongly resembles the crest of a cockerel, and may well have led to legends of crested-serpents - both in Africa and elsewhere - via visiting foreigners.

An unknown species of African snake bears an uncanny resemblance to the basilisk. The crested crowing cobra is reported from central and southern Africa. This reptile is said to be six metres (twenty feet) long, and grey or brown in colour. It has a scarlet crest like a rooster's comb on its head, as well as a pair of red wattles. Its cockerel-like attributes do not end there. The creature makes a noise very like a cockerel's crowing, hence its name. It is said to be arboreal and highly venomous. Hyraxes seem to be its favoured prey. It also attacks humans, by lunging from overhanging branches and biting their faces. Some natives, when walking through forested areas, are said to carry pots of boiling water on their heads to scald the attacking creature.
Doctor J. O. Shircore obtained some remains of a crested crowing cobra in 1944. A witch-doctor in Malawi gave him a plate of bone from the crest with skin still attached, and several vertebrae and neck bones from at least two specimens. He describes the plate thus:

"Its skeleton consists of a thin lanceolate plate of bone (1½ ins by ½ in wide, at its broadest part) with a markedly rounded smooth ridge ⅛ of an inch wide, slightly overhanging both sides of the upper border, with a distinct voluted curve to the left. The lower border is sharp-edged and faintly ridged. The lateral surfaces are concave throughout the long diameter. The whole fragment is eminently constructed for the insertion and attachment of muscles - much the same as the structure of the breastbone of a bird. Some skin, part of which spread smoothly above the plate on one side, is red in colour, and attached to the lower angle is a dark wrinkled bit, which appears to be a remnant of head skin - all of which should be valuable for purposes of identification. A small portion of bone, tapering towards both ends, ½ inch long by ⅛ inch wide, is missing from the lower anterior border, including the tip - it was broken off for use in medicine by the witch-doctor, from whom the specimens were obtained."

Shircore, it should be noted, was a medical doctor. His note appeared in the magazine

African Affairs. The remains have never been identified and to my knowledge nobody knows of their current whereabouts.

In a 1962 letter to the publication *African Wild Li*fe, John Knott recounts his brush with an individual of what may have been the same species. In May 1959, whilst driving home from Binga in the Kariba area of Zimbabwe, (then Southern Rhodesia), he ran over a two metre long black snake. Upon investigation he discovered that the reptile had a symmetrical-crest on its head. The crest could be erected by way of five internal prop-like structures.

The crested crowing cobra seems to have a smaller counterpart in the Caribbean. The eminent Victorian naturalist Phillip H Gosse records it in The Romance of Natural History, Second Series. In 1845-46 Gosse visited Jamaica where he first heard of the creature from a respected medical man.

"...he had seen, in 1829, a serpent about four feet in length, but of unwonted thickness, dull ochre in colour with well-defined dark spots, having on its head a sort of pyramidal helmet, somewhat lobed at the summit, of a pale red hue. The animal, however, was dead, and decomposition was already setting in. He informed me that the Negroes of the district were well acquainted with it; and that they represented it as making a noise, not unlike the crowing of a cock, and being addicted to preying on poultry."

Gosse's friend Richard Hill had heard of the snake from a Spanish acquaintance on Hispaniola. It was said to inhabit the eastern regions of the island, in what is now the Dominican Republic.

"My friend's Spanish informant had seen the serpent with mandibles like a bird, with a cock's crest, with scarlet lobes or wattles; and he described its habits – perhaps from common fame rather than personal observation - as a frequenter of hen roosts, into which it would thrust its head, and deceive the young chickens by its imitative physiognomy, and its attempts to crow."

Jamaican resident Jasper Cargill offered a sovereign for any specimen of the snake, but was not successful in obtaining one. Cargill himself had seen the elusive snake some years before as Gosse records.

"...when visiting Skibo, in St George's, an estate of his father's, in descending the mountain-road, his attention was drawn to a snake of dark hue, that erected itself from some fragments of limestone rock that lay about. It was about four feet long and unusually thick bodied. His surprise was greatly increased on perceiving that it was crested, and that from the far side of its cheeks depended some red coloured flaps, like gills or wattles. After gazing at him intently for some time, with its head well erect, it drew itself in, and disappeared among the fragmentary rocks."

Cargill's son shot a specimen some years later.

"...some youngsters of the town came running to tell me of a curious snake, unlike any snake they had ever seen before, which young Cargill had shot, when out for a day's sport in the woodlands of a neighbouring penn. They described it as a serpent in all respects, but with a very curious shaped head, with wattles on each side of its jaws. After taking it in hand and looking at it, they placed it in a hollow tree, intending to return for it when they should be coming home, but they had strolled from the place so far that it was inconvenient to retrace their steps when wearied with rambling.

When the youths returned the next day, the corpse was missing - presumably taken by some scavenger. When the tale was recounted to Richard Hill, his godson Ulick Ramsay, told him that he too had seen such a snake shortly before:

... not long previously, he had seen in the hand of the barrack-master-sergeant at the barracks of a Spanish town, a curious snake, which he, too, had shot among the rocks of a little line of eminences near the railway, about two miles out, called Craigallechie. It was a serpent with a curious shaped head, and projections on each side, which he likened to the fins of an eel, but said were close up to the jaws".

The legendary basilisk has lent its name to another deadly snake. One of the first venomous snakes met by European settlers in North America was the Mexican west-coast rattlesnake. Due to its potent poison its scientific name is *Crotalus basiliscus*.

Often, it seems the fear of snakes leads them to be blamed for all sorts of baleful happenings. In Rome, during the reign of Pope Leo X (1475-1521 Pope: 1513-21), a basilisk was captured, and blamed for an outbreak of plague. Another was said to lurk in a well in Vienna, and killed people with its pestilent breath. It was discovered in 1202. In 1587 a specimen was killed in a cellar in Warsaw after causing the death of several locals. It turned out to be a disappointingly small snake. The others were also probably harmless snakes found in areas where sulphur or methane fumes had built up to dangerous levels, or where natural diseases had broken out. (Remember the wyvern's propensity for spreading disease).

But what of the other factors in the basilisk legend? It seems that these also have explanations within the realm of the natural rather than the supernatural. The miraculous egg-laying cockerel is not so fantastic as it at first sounds. There is a disease in fowl that causes a hen's ovaries to become infected. This prevents the production of the female hormone oestrogen. Oestrogen controls feminine characteristics, and when these are prevented from developing, masculine traits appear. These include developing a comb and wattle, crowing, and attempting to mount hens. If the victim recovers, it returns to its former feminine self and may lay once more; ergo a cock that lays eggs.

How about the snakes that sometimes slithered out of hens eggs, to the mortification and horror of medieval cooks? Snake eggs are leathery-shelled and not at all like bird-eggs, so confusion between the two - or deliberate mischief (a jester switching hen eggs for snake eggs for example) - is unlikely. The explanation is almost as grim as the original legend! Chickens often suffer from round-worms (*Ascaris*); endoparasitic, internal creatures that are mainly passed out in the bird's droppings but they can on occasion enter the reproductive-tract and be incorporated into an egg. In times past - when there were no stringent hygiene laws - this would have occurred far more often than today. Round worms can measure up to forty centimetres (sixteen inches) and one could readily imagine the terror evoked by cracking open an egg to find a writhing "basilisk" within!

Not all crested snakes are small like the basilisk. Some are titanic. Swedish scientist Gunnar Olof Hylten-Cavallius published a book on giant snakes in his country in 1885. In *On the Dragon, Also Called the Lindorm* he published forty-eight verbatim accounts, half involving multiple witnesses. He writes:

"In Varend (in southern Sweden) - and probably in other parts of Sweden - a species of giant snakes, called dragons or lindorms, continues to exist. Usually the Lindorm is about 10 feet long but specimens of 18 or 20 feet have been observed. His body is as thick as a man's thigh; his colour is black with a yellow - flamed belly. Old specimens wear on their necks an integument of long hair or scales, frequently likened to a horse's mane. He has a flat, round or squared head, a divided tongue, and a mouth full of white, shining teeth. His eyes are large and saucer-shaped with a frightfully wild and sparkling stare. His tail is short and stubby, and the general shape of the creature is heavy and unwieldy."

He writes on the creature's hypnotic powers, its method of attacking prey, its terrifying effect on human witnesses, its hardness to kill and the foul stench of dead lindorms.

In England there are a variety of dragons. Most have wings, four legs and spit fire but some look like giant snakes. The worm of Handale was one of these. It was a giant snake with a crest and a sting in its tail. It made a habit of eating maidens until a youth called Scraw managed to slay it by pushing a sword between its jaws. He found a live girl in its cave. For rescuing the girl his reward was her hand in marriage and vast estates. The woods near Handale priory are called Scraw woods to this day.

Ovid in *Metamorphoes* book III describes the dragon slain by Cadums has having a crest, a sting, constricting coils, and venomous breath.

Philostratus stated that mountain dragons had crests that grew larger the older the creatures became.

In TH White's *Book of Beasts,* a translation of a medieval bestiary, the dragon is said to have a crest or crown and that he is the 'king of pride'. It is also listed a serpent and thought to have been large enough to constrict full grown elephants.

In the 1840s a stone coffin with an inscription so worn it was unreadable was discovered in the ruins of Handale Priory. The coffin contained an Elizabethan sword. Tradition had it that it was the grave and weapon of Scraw.

A creature with abilities akin to the Ninki-Nanka is found in the swamps of the Sudan. It is know as the Lau. Natives describe the beast as twelve-thirty metres (forty to a hundred feet) long, thick about as a donkey and yellow in colour. Some descriptions furnish it with a crest or mane, (a curious appendage for a snake but one seen in several areas). Strangely, it is also said to possess facial-tentacles with which it grabs its prey. Another reoccurring motif is horns or tentacles on the head.

The folklore attached to this monster is singularly bizarre. If the lau sees a human before he sees it, the man will die. Conversely, if the man sees the lau first it will be the serpent that expires.

The 1920's explorer and naturalist J. G. Millans interviewed a westerner who firmly believed in the monster. Sergeant Stephens, (who was never identified with a first name), told him, *"One Abrahim Mohamed, in the employ of the company (a telegraph company), saw a lau killed near Raub , at a village called Bogga. The man I knew and closely questioned. He always repeated the same description of the monstrous reptile. More recently one was killed by some Shilluks at Koro-a-ta beyond Jebel-Zeraf (Addar Swamps). I obtained some of the neck bones of this example from a Shilluk who was wearing them as a charm. These I sent to Deputy-Governor Jackson (now of Dongola province), who, in turn, sent them to the British Museum for identification, but no satisfactory explanation was given, nor was it suggested what species of snake they could belong to".*

Abrahim's story of the size and shape of the great reptile was corroborated by one Rabha Ringbi, a Nian-Niam from the neighbourhood of Wau in the Bahr-el-Ghazal, who had seen a similar monster killed in swamps near that place: "Dinkas living at Kilo (a telegraph station on the Zeraf) told me that the lau frequents the great swamp in the neighbourhood of that station and they occasionally hear its loud booming cry at night.

"A short time ago I met a Belgian administrator at Rejaf. He had just come back from the Congo, and said he was convinced of the existence of the lau as he had seen one of these great serpents in a swamp and fired at it several times, but his bullets had no effect. He also stated that the monster made a huge trail in the swamp as it moved into deeper water."

Another intriguing piece of evidence was photographed by Captain William Hutchins and published in the magazine Discovery. This was a wooden ritual-mask of the beast. When Hitchins questioned Meshengu she Gunda, the native singer and sculptor who made the mask, as to the beast's existence, the African replied philosophically: *"I might have said, as a young man, when I was ignorant, that there was no such thing as a motor car. I had never seen or even heard of one then. But there is your motor car in the sight of my eyes and I have sat on its chairs and heard its bowels digest inside it. It is thus of the lau".*

As far as I know, there have been no recent reports of the lau. Perhaps this is unsurprising given Sudan's recent troubled times. The facial tentacles mentioned recall an aquatic Asian species *Epeton tentaculaatum*, the fishing snake. Could the lau be a giant African analogue? No-one will know until the trackless-swamps of the Sudan are once again penetrated.

Captain C. R. S. Pitman - a British naturalist and expert on African snakes - was told of a titanic serpent that had inhabited a pool in the Bwamba escarpment in Uganda. It was said to be of venerable age and was worshipped by the locals. In a strange turn-about, Pitman (or his source), never explained why the snake was killed and eaten. Perhaps the animal died of natural causes and the tribesmen took advantage of the flesh, or maybe it had killed livestock or people. In any case, Pitman was told of the creature's massive size:

His informant said, *"I have no reason to disbelieve what the head man says: he is a reliable man and he measured it with a linear tape measure... every single soul who was present when the snake was measured states that it much exceeded the tape measure... I was shown the place where they stretched it out... This was certainly approximately 130 feet."*

Pitman's informant obtained the jaws, teeth, and two of the vertebra. Pitman himself never saw these relics and they were never examined by a zoologist. The informant may have been telling a tall - or in this case long - tale. His "certainly, approximately" statement does not inspire confidence. I would be inclined to dismiss this story if it were not for the other sightings of giant snakes in Africa.

Further west, something very like the lau may have been photographed by a Belgian military helicopter-pilot in the Katanga region of Zaire in 1959. According to William Corliss, the photographs were apparently taken at a low altitude, and purport to show an unbelievably colossal snake entering a hole by some termite mounds. The reptile is pictured so clearly that even the scales on its hide are visible. The photographer was one Colonel Remy van Lierde, who claimed that the snake reared up at his helicopter. The original pictures were submitted to the Eighth Reconnaissance Technical Squadron U.S Air Force experts in Massachusetts for analysis. The vegetation surrounding the snake may have been unassuming shrubs or giant trees, (I assume no botanists have ever examined these shots), but also in the picture are several termite mounds. These colonial insects build concrete hard nests six metres (twenty feet) tall as par for the course. But are these old, huge, nests or small new colonies? The termites build their nests so that the largest face receives the most sunlight in early morning and afternoon. Judging by the intensely cast shadows, the Air Force technicians estimated the termite mounds to be around six metres tall. This meant that the surrounding trees were twelve to fifteen metres (forty to fifty feet) tall. This would make the serpent a full sixty metres (two-hundred feet) long!

However, having examined the photographs, I can see no signs of termite mounds and nothing else that gives any sense of scale. Remy van Lierde is an unusual name, and as he had risen to the rank of Colonel by 1959, one can assume that he may well have been the same man who as a Flight Lieutenant flew a Hawker Tempest fighter-plane out of Newchurch in 1944. If we assume that he was a fighter pilot – and we

This is a reconstruction of what the Ninki-Nanka may look like by Mark North

It is based on native and eyewitness descriptions of the creature. Most seem to suggest that it is a vast snake rather than a giant lizard. The creature far exceeds any known living species of snake. There was, however a gigantic snake that lived in Argentine some 50 million years ago. Known as *Chubitophis grandis* it was reckoned to be around 100 feet long in life. Could Ninki-Nanka and related phenomena such as the Asian Naga be descendents of this South American giant?

know he was flying a helicopter over Katanga – then we can deduce that he was flying an anti-terrorist mission, probably accompanied by a colleague with a machine gun. If this is so then he would have been flying at a relatively low altitude of 50-100 feet, and the aforesaid photograph becomes even less impressive – most probably only a large African rock python.

The deserts of northern Africa would not seem a likely place for a gigantic snake to exist but there are reports, both ancient and modern of vast, crested, snakes in both Algeria and Tunisia.

Around 250 BC at the time of the first Punic War (264-241 BC), Rome was embroiled in a prolonged struggle with the city of Carthage (where modern day Tunis stands) over the control of Sicily. General Marcus Atilius Regulus [1] led his army towards the city when he came to the River Baradas. A titanic serpent rose up from the reed beds. The men fell back in horror, and after some consultation decided to cross the river further up stream. But as the soldiers began to ford the waters, the monster reappeared and seized a man. As each of the warriors tried to cross he was snatched by the beast's massive jaws, encircled in its coils and dragged under.

After many men were lost in this way - and it seemed as if the serpent would defeat the entire army - Regulus ordered that the giant snake should be bombarded with ballistae - giant, rock-hurling catapults. Ballistae were trained on the creature, and it gradually began to retreat under the bombardment. One boulder struck the giant's skull with fatal force and the monster snake collapsed onto the bank.

The soldiers dragged the vast corpse onto the bank and measured it. It was an astounding thirty-six metres (120 feet)! The jaws and skin were sent back to Rome as a trophy, where it was on public display in a temple on Capitol Hill until 133 BC, when it was lost during the Numantine war with the Iberian Celts. Regulus himself was granted an ovation.

Time and again the priceless cadavers of cryptids are lost to science. In the case of the giant snakes of the Sahara this has happened several times.

- Africanus Leo (Hasan ibn Muhammad al Wazzan al Fasi), was a traveller and writer born in Granada in 1485. He was enslaved by European pirates but freed by Pope Leo X. He travelled widely in Africa and visited Timbuktu twice. His were the first descriptions of the city to find their way to Europe. He wrote of huge, venomous dragons inhabiting the Atlas Mountains in North Africa. Events in the 20th century may support his claims.

- In 1958, Belkhouriss Abd el-Khader - an Algerian who served in the French army at Beni Ounif, Algeria - was attacked and bitten by a thirteen metre (forty-three foot) snake. The snake was killed and its skin preserved, although it has since been lost.

- The following year a fantastical story - a sequel to Regulus's adventure - occurred near a garrison in Ain Sefra, Algeria. A monster-snake that had just swallowed a whole camel was captured in a trench that had been filled with branches by nomads. A French battalion - the Twenty-sixth Dragoons - were brought in to kill it. Their carbine rifles did little damage and they had to finish the monster with machine guns. The beast was thirty-six metres (120 feet), the same size as Regulus's reptile, and bore a metre long (three foot) crest on its head. No-one seems to know what became of the body.

- In early January 1967, a nine metre (thirty foot) snake was seen on the construction site of the Djorf-Torba dam, east of Bechar, Algeria. A worker called Hamaza Rhamani wedged it against some rocks with his bulldozer. He reported that the beast's fangs were some 2.5 inches long. Later that year, in the same area Rhamani saw another specimen. He followed its trail to some barrels of oil from which - bizarrely - it seemed to have been drinking. He saw the snake coiled in the shadow of a pile of crushed rocks and estimated the length to be five-seven metres (eighteen to twenty-three feet).

But it is the Asian Naga that is the best known of all giant crested snakes. Its wonderous image can be seen gracing temples from India to Korea.

In 1966 a peasant digging in the mud on the bank of the Mekong, close to the Lan Xang hotel in Vientiane, the capital of Laos, uncovered some huge white eggs. Subsequently he claimed that the Naga appeared to him in a dream, demanding the return of its eggs, and threatening to flood the river if they were not given back. He took the eggs to General Kong Le, leader of the Neutralist army. The general showed them to Premier Prince Souvana Phouma, and warned him of the impending peril. Le was to lead the people in a ceremony of atonement but the Prince was unimpressed by the peasant fairy tales. The ceremony was never carried out, and the monsoon rains brought a huge flood to the city. Tragically, no-one knows what happened to the Naga eggs.

Back in Thailand, a strikingly similar event was recorded in May 1980. Fisherman Prancha Pongpaew found seventeen eggs in the River Ping, north Thailand. The eggs were the size of water-melons and seemed to be linked together in a chain like toadspawn. He brought five of them to the surface and took them back to Songhtam village. The eggs were broken but they smelled so bad that they were thrown to dogs who ran away in fear.

That night, the villagers were awoken by an odd wailing sound and were horrified to see two black serpents - the size of palm trees - with crested heads, rearing up out of the river. The following night, a religious ceremony was performed on the banks of the river but the Nagas did not reappear.

---

1. In the early 1990s, a team of Japanese cryptozoologists claimed to have taken film of an unknown aquatic creature in Lake Dakataua, on the island of New Britain, in the Bismarck Archepelago, near New Guinea. Cryptozoologists around the world were excited by this news, but it was the first big break for the CFZ when we managed to get hold of a copy. Our friend and colleague Darren Naish identified it as a pair of Indopacific crocodiles, probably mating.

It should be noted that snakes either lay eggs on land or give birth to live young. The strand of linked eggs in water is very odd and distinctly un-reptilian. They sound like outsized toad spawn. Perhaps the Nagas and the eggs were two unconnected events.

The year previously, a Naga had caused a stir in Malaysia when it turned up in a disused mining pool in Semenyih. Fisherman Lebai Ramli saw it rise up from the water, and fled in terror. The incident caused a local stir as crowds of people, some armed, swarmed to the pool to try and see the monster. Sign-boards were put up telling people the way to the pool, and there was even an ice cream man on hand! Amazingly the Naga obliged. It surfaced at about 12.30, when many witnesses saw a creature with a head the size of a scooter-wheel. Farmer Enick Arshad described seeing a log-like creature swimming 12-20 feet from the bank.

Enick Jaafar's 12-year-old son claimed to have seen a snake-like animal with a head - the size of an oil drum - held two-feet above the water. Enick himself saw the creature shortly after World War II. He described it as a snake the size of a tree trunk.

The monster was held responsible for the disappearance of two buffaloes and other local livestock.

I have hunted for the Naga on a previous CFZ expedition so I present the expedition write-up here in its entirety.

As Britain's only full-time cryptozoologists, we at the CFZ are frequently contacted by researchers from TV companies wanting to make monster-related programmes. We are habitually visited by bright eyed young media things, who get incredibly excited and tell us that our ideas would make excellent documentaries. They then, invariably, disappear for good. In the worst cases, we find our ideas stolen and bastardised by people who know nothing about the subject.

Therefore, when we were contacted by a company called Bang Productions in July of 2000 we did not hold any great expectations. We were visited by an outlandishly beautiful half-Japanese girl called Manami Szymko, who had come all the way from Hong Kong (where the company was based), to interview us as possible presenters on a Discovery Channel project called Asia Enigmas. In particular she was interested in the Naga, a gigantic legendary snake reputed to inhabit the Mekong River. Other episodes will feature the yeti, ghost hunters in the Philippines, Indian holy men with healing powers, the supposed undersea city off Japan, and UFOs in China.

Manami interviewed me at length about the monster and filmed a screen test. She met all our exotic pets and then vanished and we never expected to hear from her again. That was in July 2000. Imagine then, gentle reader, your humble narrator's surprise when a young lady, Sandra Egart from the aforementioned company, called in early October asking if I could join them in Thailand in a handful of days' time.

The next few days were a blur of injections, and procurement of tropical accoutrements. Then all of a sudden I was thousands of feet over Asia on my way to *Boy's Own Paper* style adventures.

It may be prudent at this point to give background on the Naga itself. Nagas are gigantic snakes found in Hindu and Buddhist mythology. They bear an erectile crest upon the head like that of a cockatoo, but consisting of scales. The Naga holds this aloft when angry rather like a cobra opens its hood. According to Buddhist scriptures, the Naga can kill in four ways. Firstly by biting and injecting its venom and secondly by spitting like certain species of cobra. In this case the venom has a paralysing effect causing the victim to become as stiff as a statue in death. Thirdly, by constriction with its powerful coils, and fourthly by its baleful glare, much like the basilisk of medieval Europe and the Middle East.

According to legend Nagas have immense intelligence and magickal power. They could transform themselves into humans and walk un-noticed in the world of men. It was believed they inhabited grand underwater palaces rather like the dragons of China. Unfortunately for folklorists of the Michael Meurger ilk, the Naga is not satisfied with being a legend and still rears its scaly head today.

The flight from England via Amsterdam took an uncomfortable ten hours, but finally I arrived in Bangkok. Sandra, the production assistant who had contacted me earlier, and Peter Daniel, the producer, met me at the airport. I was surprised at their youth, having expected middle-aged people. Sandra, a former model, was of a particularly striking beauty.

I had been told that due to budget restrictions we were staying in a cheap hotel. "Cheap hotel" seems to have a different meaning in Thailand. The Amari Atrium, in which we stayed whilst in Bangkok, was by far the finest hotel I have ever had the pleasure of patronising. This begs the question of what an expensive hotel would be like.

Presently, the other members of our crew joined us. The researcher and interpreter, Athihan Srivetbodee, or "Bob" for short, who also worked for a charity protecting captive elephants. The cameraman was Derek Williams who, in a thirty-year career, had covered just about every event of importance in Indo-China. His mother has been badgering him to write his autobiography for years. I, for one, would love to read it. He was ably assisted by his soundman, Somyot Pisapark, who had accompanied him on numerous previous adventures. Somyot was a dedicated man. Halfway through our filming schedule, he was told his wife had developed throat cancer, but he continued all the same.

Bangkok is a strange city. It bares an uncanny resemblance to Birmingham. It even has an office block shaped like Birmingham's famous Rotunda. Gaining planning permission in Bangkok is as easy as fancying Kim Director. So buildings spring up like fungi. So fast do they get built that often times some small things like foundations or strengthening rods are forgotten and the building is abandoned. Unlike Birmingham, however, one often comes across an elephant wandering nonchalantly down the street or rooting through a bin outside a bakery!

Later that evening, I was shown some film of the giant Mekong catfish (*Pangasianodon gigas*). This animal is the largest (in terms of bulk) freshwater fish in the world and has been mooted as an explanation for the Naga. The sequence showed four men catching an eight-foot specimen. The silvery grey fish is of massive bulk and has bizarrely situated eyes, very low on the head. The men manually stimulated the fish's cloaca to collect its milt to use in captive breeding programmes. Strange to think I had travelled all the way to Thailand to watch film of a fish being masturbated!

The day after my arrival, we visited Samutprakarn crocodile farm , home to the largest crocodile in captivity: a 20 foot Indo-Pacific, Siamese cross named Yai. Yai in Thai means big , what a lot of thought went into his naming! Yai was sharing a network of pools with around 100 other crocodiles. Conveniently for us, he was in a small shallow pool that allowed me to walk up and down his entire length and confirm his size.

The keepers swore that Yai was the largest crocodile on the farm, but out in the main lake I saw a number that appeared to be several feet longer. Two specimens looked around 23 feet, and a huge individual appeared to be around 25 feet. This latter giant stayed in the centre of the large pool and would not be tempted closer to the bank. He showed only the end of his huge jaws and a portion of his scutted, tree-trunk like tail. Ergo an accurate measurement could not be made.

I had a theory that the Mekong monster could be a large (30 foot +) Indo-Pacific crocodile *(C. porosus)*. I later abandoned that idea after hearing eye-witness accounts, but this gave me a chance to view my favourite creatures closer up than ever. The crew had me talk about crocs and the titanic sizes they can reach, whilst filming me in front of the pool. Then the gates were opened and I was presented with several buckets of chicken carcasses. "You lean through and feed them. We'll film you from over here," they said.

My days as a zookeeper taught me that captive crocodiles are much more interested in eating the food a keeper presents them with than the keeper himself. Yai was not hungry but several of his comrades came whizzing in like Polaris missiles with bear traps attached to them. I must admit to enjoying feeding them immensely and became nostalgic for my years as a zookeeper.

Samutprakarn would not past muster for a zoo in the west. Its promotional material pushes the conservation angle, but by cross breeding (the Indo-Pacific's huge size and fast growth with the Siamese's less aggressive nature being the 'ideal mix' for skin farming) the gene pools of both species are being diluted. The Siamese crocodile *(C.siamensis)* was, until recently, believed to be extinct in the wild, but thankfully they seemed to have survived unscathed in Cambodia where they were rediscovered only this year!

Elsewhere at Samutprakarn, were tigers attached to four foot chains that visitors could be photographed with. The docile creatures appeared to have been doped. Baby orang-utans were paraded in dresses for the same purposes.

This day seemed to be one for appalling zoos, because in the afternoon we visited

Pata Zoo in downtown Bangkok. Believe it or not, this zoo is situated atop a department store. One of the floors houses a reptile collection that is not badly maintained. Also here they had a preserved specimen of a creature I had only ever read about in Karl Shuker's *Lost Ark*, the giant freshwater stingray (*Himantura chaophraya*). Only discovered in 1987, this fish is a monster in every sense of the word.

The nearest description of this piscine titan I can give, is an organic flying saucer! Greeny-grey in colour, its flattened body disc measures some seven-feet by six and a half feet, big enough to cover a double bed. Its eyes were tiny like those of the Mekong catfish with whom it shares its habitat. These bottom feeders must rely on touch, scent and electro-reception in the Mekong's muddy waters.

On the roof of the building, the mammals and birds are kept in truly appalling conditions. I am an ardent supporter of responsible zoos with good breeding programmes - they are vital to save many endangered species. But slapdash holes like Pata Zoo belong in the dark ages. Here gorillas, tigers, leopards, orang-utans, and pigmy hippos were kept in enclosures the size of the average living room. Worst were the bears. Three sun bears and an Asian black bear in a bare concrete enclosure, with no den or climbing facilities. It could not have been more than ten-feet square.

A woman was selling cakes to feed to them, so their existence was nothing more than sleeping and begging. Ironically, the zoo had some rare animals hardly ever seen in British zoos such as umbrella birds, Burmese ferret badgers, and yellow martens. These were totally wasted as exhibits in such a vile excuse for a zoo.

We were meant to be interviewing the director of Pata Zoo, who had taken some film of an alleged Naga swimming in the Mekong. However, he had fallen over and banged his head and was in a critical condition in hospital at the time. It seems karma really works! Instead, we talked to a Dr Apicsart, who was a fish expert not connected to the zoo. Dr Apicsart had spent many years on board Japanese trawlers studying rare fish, often from the deep seas. He was sceptical about the Naga, believing witnesses had seen shoaling fish. As I was to later find out, this explanation did not stand up to scrutiny.

The following day we left Bangkok and flew to Udon Thani in the north of the country, that would be our base of operations for the rest of the expedition. We were met at the airport by the other main player in the adventure, Pongpol Adireksarn, better known in the west under his pen name Paul Adirex. A best-selling author, both in east and west, he specialises in action thrillers and has penned four best sellers one of which, *Mekong*, features Nagas, in which he firmly believes.

As well as this, Pongpol is the campaign manager of the opposition party in Thailand. Charming and witty, he was a joy to work with. When I asked him if his politics were left or right he answered, *"In Thai politics there is no left or right, just right and wrong!"*

In Udon Thani we checked into our hotel and were met by some of Pongpol's friends and assistants, including a friendly and enthusiastic young man called Pracha Manakarn - or "Pang" to his friends. Pang owned a pub in Udon Thani and wanted to become a tourist guide. I never did find out exactly what he did for Pongpol, but he proved to be an excellent companion and addition to the team.

That evening, a banquet in Pongpol's honour was held at a local restaurant. He seemed to be a very well-liked man (unlike most western politicians), and a local celebrity. It was here that I ate Mekong catfish soup. I'm not a great lover of fish (as food that is), but the Mekong catfish had some of the most delicious flesh I have ever eaten. If you can imagine the most succulent melt-in-the-mouth fish mixed with Parma violets you might get some idea.

The following day we drove to a statue garden. Here, gigantic statues some 150 feet or more tall, stand, surrounded by jungle. One could be forgiven for thinking they had stumbled upon a lost city of the H. Rider-Haggard kind, but the statues are only about 20 years old.

They depicted mainly religious characters. Thailand is a Buddhist country, but their Buddhism is singular in that it has been heavily influenced by Hinduism. Hence, alongside sculptures of the Lord Buddha, Gods such as Kali, Ganeesh, and Hanuman are found. This is an important factor that I shall return to later.

There were several massive statues of Nagas including the multi-headed Naga king shading Buddha as he meditates.

As well as the Naga, other Thai monsters were present. The garuda is a creature half-bird, half-man, who is said to bring the rains on his wings. There is also a magickal golden lion. All three live together in a mystic jungle. So, we have a giant water serpent, a bird man, and a mystery big cat in the same country. Sound familiar? It seems there are universal templates for monsters. One could just as easily be talking about Morgawr, Owlman, and the Beast of Bodmin.

Later in the town of Nongkhai, we spoke to Malinee Phisaphan, an old lady who had seen a Naga five years earlier. Malinee ran an antique stamp shop and a cyber café! She was highly intelligent and read, wrote, and spoke perfect English. She and a friend had been riding through town on a bus when they passed by a bridge.

Both of them saw a huge black snake in the water beneath the bridge. Malinee described it as around seventeen feet long and as thick as a football (that is an English football). It could have been a large python but it would have had to have been melanistic. The next day was an important one. The Naga mystery has been mixed up with several other mysteries in its complicated folklore. One of these is the Naga fireballs.

On October 13th every year, balls of red light are seen shooting out of the Mekong River. Locals believe these to be the breath of the Naga and to herald the end of the rainy season. Huge crowds are drawn to celebrate and view the phenomena from the banks of the river.

The Fortean in me noted the balls of blue light associated with giant snakes in the

Amazon and believed to be their bioluminescent eyes. I also thought of earth lights often reported over water. Perhaps two Fortean phenomena were occurring here side by side?

In the daytime, before the nocturnal lights appeared, a huge parade was run. This included hundreds of people in traditional dress, bands, and floats holding images of the Naga. It ended with a temple made entirely from bamboo leaves being floated upon the river.

I am a zoologist, but I am also a pagan, a practising magician, and a believer in magick. I had prepared my own offering for the Naga that I had brought all the way from England. It consisted of a piece of jade (oriental dragons are fond of jade according to legend), incense, and a resin statue of a dragon.

I was planning to float this down the river as an offering. Unfortunately, I had not realised the steepness of the bank or the thickness of the crowd that night and had to postpone my offering's launch.

I found myself surrounded by 100,000 screaming Thais, shining spotlights and laser pointers on the water and letting off fireworks. Traditional long boats, illuminated with candles and lamps, passed by as we waited for the phenomena to begin. Suddenly, a shout went up, a fireball had been spotted. Shortly after, I saw a red light spring upwards from the opposite bank then fade away. Soon, more followed; firstly in the singular then in twos, threes, and fours. Swiftly something dawned upon me - if this were a natural phenomena it would be occurring from the entire width of the river. All the lights were springing up from the far bank of the Mekong, i.e., the Laotian side. Also, they were all coming from areas were lamps were visible and presumably people were present. They also looked very orchestrated. The fabled Naga fireballs seemed to be nothing more than fireworks, the relatively noiseless kind that fade away rather than exploding, much like maritime distress flares. I mentioned this to some of my Thai companions, who said they couldn't possibly be fireworks because they were silent and faded rather than exploding!

You have doubtless heard of the Loch Ness "hoodoo", the ill-luck that befalls those trying to photograph the monster. Cameras jam or are forgotten, or something else happens at the crucial moment to hamper the picture. Well, I suffered from Mekong hoodoo. Earlier that day I had checked the charge on the batteries of my video camera. They said I had over an hour of film left.

When I tried to film the festival the camera gave out halfway through the pre-fireball parade (but if you're here who is grooming the Nagas ready for the Naga parade?). Hence I had to rely on a stills camera.

A couple of days later, as I tried to rewind, the back opened and all the film spooled out. I wouldn't read too much into this as I am the most cack-handed and incompetent technophobe you are ever likely to meet. So the fireballs bit the dust. I was satisfied that the Laotians were having a good chuckle at their friends across the river, but there were other riddles waiting for my attention. The next day, we visited the village of Phon Pisai were Pongpol interviewed a Buddhist abbot (who bore an uncanny resemblance to the late Brian Glover), and his monks about a strange Naga encounter. The temple was the most spectacular building in the village, adorned with dragons and Nagas.

Eight years ago, there was an old temple where the fine new one stands. The ancient, tumbledown erection had become unsightly and dangerous so it was decided to pull it down and build a new one, but whenever workmen approached, a huge black snake would appear and rear up striking at them.

Workmen, monks, and the abbot all saw it. It was very thick, but they could not estimate the length as the creature never revealed its whole body, but kept most of its coils in the building. Finally, an offering was given to the monster and it disappeared overnight.

Back in Nong Khai we interviewed a Dr Manus. He had a theory on the fireballs. According to him, on October the 13th, the Earth is in such an alignment with the Sun that the solar waves are at a certain length that in some way effects gas molecules in the river and energises them into fireballs. He had some gizmo for reading gaseous emissions and showed us that around the 13th, and a couple of days either side, the emissions rose. I'm no physicist, but all this sounded a little dodgy to me. As it turned out Dr Manus was no physicist either, he was a paediatrician! He gave me the address of his website, but I lost it on my way home and have not been able to locate his site on the net. If any readers have better luck please let me know.

The following day, we were back in Phon Pisai to interview another witness. Officer Suphat is chief of police in Phon Piasi. Three years ago, he and a group of thirty people had been walking on some cliffs overlooking the Mekong.

They had seen what, at first, they believed to be flotsam floating along in the river. As it drew closer they became aware that it was moving against the current.

Looking down they saw a gigantic black snake swimming with a horizontal flexation (indicative of a fish, amphibian, or reptile).

I asked Officer Suphat how long the monster was. His answer staggered me, seventy metres! I double-checked thinking there had been a mistranslation, but he clarified seventy metres or two hundred and thirty feet. A monster of truly Toho Studios dimensions.

The crowd watched as the Naga swam by, then were overcome by fear and fled. He later asked a Buddhist monk about his sighting. The holy man confirmed what he had seen was a Naga. He explained that some years ago, a statue of Buddha was being transported by boat across the river. The boat capsized and the statue fell to the river bed. Since then Nagas have come to protect it.

The officer's monster seems excessively long. I think what he may have seen were several Nagas swimming in line, perhaps males in pursuit of a female much as the

"Migo" footage from New Britain [1] shows two crocodiles swimming in single file. Alternatively, it could have been a long wake that made an already huge serpent seem even longer.

Another enigma awaited me in Phon Pisai, one that excited me as a zoologist. The bones of an actual Naga were said to be kept as holy relics in the village! A strange story was attached to them. Their current owner had a dream in which he was visited by a Naga. The serpent told him to cross the bridge into Laos where he would meet a man who owned Naga bones. He was told to buy the bones. Duly, the man crossed the bridge the next day and met the bone owner. But the Laotian refused to sell the bones and the Thai returned empty handed. Next night, the Naga came to his dreams again and told the man to return and ask the Laotian if he would sell half the Naga bones. Once more the bones' owner refused and the man came home with nothing. One final time the serpent entered the man's sleeping mind and told him to try once more and that the Laotian would relent. The man crossed the bridge a third time and indeed the other conceded and sold him the bones.

This aside, I was excited at the prospect of laying my hands on real physical evidence of the creature. We were told that the owner did not want to be filmed and would not let us take the bones away for DNA analysis as I had wanted. However, we were allowed to film and touch them. I was confident about being able to identify snake bones and hoped we had stumbled across evidence of a titanic new species. The bones were brought to the police station and kept under lock and key until we arrived. They were brought out in a silver chalice. We waited with baited breath as the lid was removed to reveal a sodding elephant's tooth! Quite how, in a country so jam-packed with pachyderms anyone could mistake an elephant's tooth for anything else is beyond your humble narrator. Another mystery shot down in flames.

My final full day in Thailand turned out to be the most exciting and fruitful. We drove for hours north along dirt tracks in the jungle, then trekked on foot to an extremely remote village in the forested hills. I don't even know if this place has a name, I never found out. As Rangi from *It Ain't Half Hot Mum* said: "It's the back of beyond and then some". Our witness was a sprightly old man of about seventy, called Mr Pimpa.

He had a frighteningly close encounter with a Naga in some little known caves in the area. After filming and interviewing Mr Pimpa, he offered to take us underground to the Naga's lair. We were led to a small cave mouth in a hillside hidden by the verdant morass. It did not look like much, but Mr Pimpa told us that it led to a network of caves that stretched for some ten miles beneath the hills and connected with the Mekong. Like a guardian, a strangely flattened and cryptically-coloured spider four-inches across, lurked on the lichen at the cave mouth.

The camera crew filmed the entrance and Peter, the producer, followed Mr Pimpa, Pang, and I into the first cave. Roughly fifteen-feet square and four-feet high, it did not look like much, but by flickering candlelight, our aged guide showed us a tiny triangular tunnel in one corner. Peter, with his expensive hand-held camera went no further and I left my bulky camera behind too, as Pang and I followed Mr Pimpa.

The tunnel was half filled with water and so low one had to crouch. It led on for some forty-feet into the main network. It was as if we had entered the fevered mind of Clark Ashton Smith. These caves were, by far, the strangest and most alien place I have ever been in. Imagine a hybrid of the labyrinth of Sogo in "Barbarella" and the "Caves of Androzani" from the eponymous Dr Who adventure. Now shrink them. None of this honeycomb was more that four-feet high. The dank, unwholesome passages were usually half that wide. Occasionally, they widened out into spaces of fifteen feet. These were peopled with unearthly rock formations like giant coffins or Greek pillars. All were festooned with offerings of jasmine wreaths in honour of the great serpent.

On several occasions we had to cross icy subterranean rivers and navigate razor-sharp stalactites that hung like guillotines from the ceiling. When not crouched, we were on all fours or slithering like worms on our bellies through the primal slime. No bats hung from the ceiling, but I observed what looked like tiny glowing strings of pearls hanging from the cave ceiling. These were drops of luminous saliva suspended on strands of silk by carnivorous midge lava like ghoulish fishing rods. I have only heard about these from caves in New Zealand and never elsewhere. Unfortunately, I was not carrying a specimen jar (a near physical impossibility down there), so I could not collect any. Does anyone out there have any idea if this is a new species?

We travelled for about a mile until we came to the place where Mr Pimpa had seen the monster some ten years ago. It was an elongate tubular cave. The old man had been exploring by candlelight, when he had turned into this cave and come across a giant snake. Its head was in the shadows but the visible portion of its body was sixty-feet long. Mr Pimpa pressed himself back against the wall in terror as the giant reptile crawled by at an agonisingly slow pace. Its scales were black with a glossy green sheen and it was around two and a half to three-feet thick. Finally, it disappeared along the passage and Mr Pimpa collapsed gasping in relief. In the dark, his had fell against a tiny semi-precious stone which he pocketed. Scrambling back out of the cave system, he returned to the village and told his weird tale. He later had the stone mounted onto a serpent shaped ring which he showed to us. He believed that, despite the fear he felt at the time, the Naga brought him luck. Prior to his adventure he was a poor man who could hardly afford to feed his family, but after it, he inherited some land and became a successful farmer. The caves were now considered sacred to the villagers.

Fortunately, the resourceful Pang had a tiny pocket camera and took shots of me in the Naga cave. He is posting them on to me shortly.

He led us back along a different set of passages and I regretted not having brought a ball of twine. Suddenly, daylight streamed in and I looked up to see a vertical shaft ten-feet high with perpendicular slime-covered walls. Mr Pimpa shot up it like a monkey, but a portly, clumsy, cryptozoologist is not the most agile of creatures. After several attempts, I was forced to climb up poor Pang like a living ladder and be dragged the rest of the way by our guide. We then pulled up trampled Pang and trekked back off through the jungle to our crew.

I was impressed by Mr Pimpa's testimony. He had nothing to gain from lying to us and was not paid for his story, and seemed genuinely surprised that people from the outside world were interested. He was a very nice man who went out of his way to be helpful.

That night, back in the hotel, I was shown the film taken by the director of Pata Zoo of the supposed Naga swimming in the Mekong. Most film of cryptids is bad, fuzzy, pixelated, out of focus, but this took the biscuit. It was a wobbly, badly filmed log being bobbed up and down by the current. Nothing more, nothing less.

We had a goodbye drink at Pang's pub, "Made in Udon Thani". It's a great place, with live bands, beautiful bar maids, good food, and good beer. If you're ever in Udon Thani be sure to check it out and give Pang my regards.

The next day we flew back to Bangkok and awaited our transport home. Bob, Derek, and Somyot returned to their abodes in Thailand. Peter, Sandra, (who live in Hong Kong) and I waited for our planes. We had a drink with an old friend of Peter`s, Mike Dyer, a computer programmer who married a beautiful Thai girl and lives in the country full time.

Sandra and Peter's flight was several hours before mine, so Mike kindly stayed and downed several pitchers of beer with me whilst I waited for mine. He told me of the idyllic life he had led, living in a shack on a beach in southern Thailand with a lovely Thai girl selling tee-shirts to tourists, until they built a hotel over his shack. At the moment his wife is very ill with a respiratory disease caught from bat guano in some caves. My best wishes go out to both her and Somyot's wife.

I slept most of the flight back, and returned to cold, rain, and floods. After the laid back attitude of Thailand, it was as if a tidal wave of woe had broken over my head, but forgetting my moaning for a moment, what conclusions did I come to?

Firstly, the fireballs seem to be man-made, possibly in order to attract custom to the area (stall-holders really cleaned up on the 13th). Secondly, the Naga bones were elephant teeth. Thirdly, the Naga film was a floating log in the Frank Searle mould.

One mystery remains unbowed, however - the Naga itself. The witnesses seem to fall into two categories, those who saw something in the river and those who saw something on land. Both, however, have mystic overtones e.g serpents guarding temples and statures or bringing good look.

Do you remember me telling you about the Hindu influence on Thai Buddhism? Well this, I think, is the key. Nagas originated in Indian legend and were brought down into Indo-China. I think all of the mystical elements of the original legendary Naga have been grafted onto a real animal, something that has always inhabited the Mekong. But what is it?

There were once a group of snakes that did reach immense sizes. These were the Madtsoiids, They first evolved in the Cretaceous period and were found worldwide.

At first believed to be giant Boids, it is now known that they were a primitive basal group of snakes. These were highly successful for such archaic beasts and flourished in some cases, such as the Australian wonambi until the end of the Pleistocene epoch, only ten thousand years ago. Some species dwarf today's anaconda. A vertebra from South America indicates a sixty-foot snake as thick about as an oil drum. Primarily aquatic, it is believed they were live bearers.

Reports from all over the tropics suggest that some species may have survived to the present day. As well as great size, all these monster snakes seem to have strange ornamentation on the head. The lau of the swamps of Sudan is said to have facial tentacles, the mano tauro, or sucuriju gigante, of the Amazon is believed to have horns, and Indo-China's Naga has a crest. Horns are not unknown on snakes: the rhinoceros viper of Africa and the horned viper of the Middle East are just two. The horns are actually modified scales. Madtsoiids killed their prey by constriction with huge muscular coils, so what of the Naga's venom? Well, having both constriction and venom would be evolutionary overkill. As far as we know, no Madtsoiids were venomous. Perhaps this is a faucet of folklore, like the harmless salamanders of Europe, which were supposed to be deadly poisonous.

So there you have it, my theory on the Naga and giant snakes worldwide.

It is only a theory and will remain so until a well-financed expedition, with a lot of time, makes a concerted effort to find a specimen. One last thing about these giant snakes: it makes you wonder about all the medieval legends of giant snakes in Britain, such as the Lampton worm and the Linton worm etc. Could there once have been a temperate hibernating species in Europe?

However, my dealings with the Naga were far from over. In 2004, whilst on my second expedition to Sumatra in search of orang-pendek (an unkown species of upright walking ape), I heard more stories of these giant reptiles.

Sahar, our guide from the 2003 expedition, casually told us that he had seen a giant snake captured by a jungle-dwelling tribe called the Kubu. We instantly recognised this as the story that had reached the British press of a 49ft (15m) long, 985 lb (450kg) python called "Fragrant Flower" . The giant reptile had reputedly been looked on as an elder by the tribe. It was alleged that Imam Darmanto, the owner of a zoo in Java, had persuaded the Kubu to part with the giant – although it had taken 65 men and the blessing of a tribal leader to capture it. The snake was transported to Java, where it was put on display and fed a diet of dogs. Unfortunately, when The Guardian sent over a reporter with a tape measure, Fragrant Flower had shrunk to 23ft (7m). It seemed that the whole story was a publicity stunt by Mr Darmanto to promote his tawdry zoo.

Sahar confirmed that the snake had been about 7m long; more interestingly, he also promised to take us to talk to the very tribe who had captured it when we returned from the lost valley, where we were hunting orang-pendek.

With us putting questions to Sahar in English, Sahar asking the translator in Indonesian, and the translator asking the Kubu in their language, we succeeded in conducted an interview.

Nylam confirmed that he and his tribe had, indeed, captured a large snake - it was a python. When asked about its length, he stated that it was 23ft (7m) long. This tallied with both Sahar's estimate and the measurements of The Guardian reporter. The snake had been sold to a man in Java. The chief said that they had caught a 26ft (8m) specimen shortly after, but had let it go back into the jungle again.

I asked if any of the Kubu had ever seen a 15-metre snake. They all agreed that they had never seen one so large. I asked how long the largest snake they had seen was, and Nylam, and several of his hunters, all said they had seen several snakes of 10m (33ft). One, in particular, had been living close to their habitations about six months ago.

Now came the strange part - all three men were adamant that these 10-metre snakes sported cow-like horns. One man had been within 5m (17ft) of one of the giant snakes and confirmed that it had horns. They also said it had a moss-like growth on its back. I asked them to draw a picture for me, but none of them could draw. I produced a quick sketch of a reticulated python to which I added horns. It met with enthusiastic nods of approval.

Stranger still, were their beliefs about these huge snakes. Once a snake reaches a very large size, it begins to get fatter and shorter. It grows four legs, each with five toes. Then it swims out to sea. I drew another picture, this time of an Indo-Pacific crocodile. The Kubu all agreed that this is what the great horned snake eventually becomes. In this form they called it a Naga. They said it was larger than the common crocodile (or buaya, meaning "rascal" in Indonesian).

The Indo-Pacific crocodile does inhabit the region and, at its extreme may reach 10 metres. This is the record length for the reticulated python as well, and it is interesting that the term Naga is used for these creatures. In India and Indo-China, Naga specifically refers to a giant crested snake, possibly an unknown species. In Indonesia, Naga means dragon and appears to be used loosely to describe any monster reptile.

As far as I know, this belief that pythons become crocodiles is unique to the Kubu. Quite where such a queer fancy springs from is hard to say. No-one seems to have studied the Kubu and their folklore.

Doubtless, there are far more stories of giant crested snakes in legend and from modern day witnesses around the world. I have, I think, only scratched the surface of this ancient riddle.

NINKI NaNka

Back in Woolsery, the children of the village became very excited by the expedition. Several of them presented us with pictures they had drawn - Ross and Greg Phillips' previous page shows their drawings of Ninki Nanka and Gambo. Overleaf, Jessica Bond shows her drawings.

Jessica Taylor-bond

gambo

Jessica Taylor-bond

APPENDIX TWO

FRESHWATER ECOLOGY OF THE GAMBIA

Simon Wolstencroft
(Editor, *Tropical World* Magazine)

To many people, ecology and fishkeeping are incompatible. The first group are vainly attempting to save and maintain wild species in their natural habitats, while the second are merely fulfilling a desire for a "moving ornament".

Sadly, this is often true, but there many expert fishkeepers and specialist societies to whom the well-being of wild fish is all-important – nowhere more so than in Africa.

My remit for this article was originally to describe some of the fish of West Africa. On beginning, however, I discovered that this would simply give tantalising glimpses of a few species, while explaining nothing about their origins, lifestyles and place in the African ecology. Instead, we shall look at the various African habitats through the eyes of one of the Continent's dominant families, the cichlids.

For fishkeeping purposes, this branch of the family is generally divided into East African, and West African species. The reason for this will become clear later.

How to describe a cichlid? They all share certain characteristics, they are among the most intelligent freshwater fish, and all exhibit some form of brood care.

They are generally, but not always, squat and muscular. They are found in most parts of Africa and South and Central America, with the northernmost species occurring in Texas.

They are mostly carnivorous, and adult size can range from two inches to two feet. Most, but not all, breed as long-term pairs. Others, such as some of the Lake Tanganyika shell-dwellers, prefer a harem system with a number of females defending micro-territories, under a dominant male.

But here we must return to our divide between west and east African cichlids. This division is questionable, to say the least, and concerns species habitat and water chemistry.

It is probable, although not entirely proven, that all cichlid ancestors were originally riverine species: That is to say that they evolved in river and stream environments, until geological events caused the most spectacular examples of accelerated evolution ever.

Some 35 million years ago, the African and Arabian plates separated, causing – over a 20 million year span – the formation of the African Rift Lakes, such as Lakes Malawi and Tanganyika. Although this may seem like a long period, it is just the blink of an eyelid in geological terms.

The lakes gradually – and we are looking at a long period of time – filled from surrounding rivers. Cichlids were trapped in these waters, which were becoming more alkaline, and they had to evolve rapidly to survive. This explains the immense diversity of cichlids in the eastern parts of the continent.

Meanwhile, what was happening further west? Very little, really. While West Africa may not offer the diversity of colour and body forms of their eastern cousins, they boast some of the most beautiful cichlids in the world.

While the East African lakes are hard and alkaline, West African rivers tend toward soft, acid, water. As the river flows, vegetation tends to add tannins which add to the acidity. Vegetation is rife here, as are dead logs – a perfect home for many cichlid species – plus an insect-rich diet.

Let's look at a typical West African cichlid, *Pelvicachromis pulcher,* known in fishkeeping circles as the "kribesnsis," taken from the former Latin name.

This fish inhabits Southern Nigeria, from the Niger River to Benin, preferring costal regions.

The natural habitat is streams, or slow-moving sections of rivers. These fish are micro-predators, living on insect larvae, small crustaceans, and the like. Naturally, this diet attracts them to vegetation-rich areas.

This species attains a length of around 10cm (4in), although females rarely grow larger than 7.5cm (3in). Unusually, females tend to be more brightly-coloured than males.

Back in the 1970s, this was a "must have" fish for serious hobbyists, and was considered difficult to breed. The secret was to provide soft, acid water. The krib also needs an enclosed area in which to spawn – a small flowerpot is generally preferred in captivity – and up to 100 eggs will be laid on the roof of the "cave".

Parental care among cichlids depends on species. Sometimes only one parent tends the brood, sometimes both. The female krib looks after the young herself and will not permit the male to enter the cave. This rule extends to other intruders, and, unless you want a fire-eating, enraged hell-kitten of a 3in fish attacking your hand, any fishkeeper is advised to keep away.

A more aggressive cichlid is the jewel fish, *Hemichromis guttatus.* This fish has a range stretching from the Ivory Coast, down to the southern Cameroons west of the Volta.

Growing to about the same size as the krib, this fish is a typical example of a split personality. For most of the time it is a peaceful, laid-back, individual which would not harm anything too small to be eaten, and could safely be invited onto a TV chat show. But when the breeding cycle starts – watch out!

At this stage, the pair become intolerant of anything, and everything, apart from their own mate. They show about as much of the milk of human kindness as the average *Big Brother* contestant and will defend their brood to the death – that death normally being meted out to the threat!

They spawn by digging out pits in the substrates, where up to 500 eggs are laid. Both parents tend the young.

We shall return to the cichlids later, but there are many more exciting fish families in Western Africa. Parallel evolution, which produced minnows in Europe and Asia, has given us tetras in South America and Africa. The Congo tetra (*Phenacogrammus interruptus*) – by no means confined to the Congo – is undoubtedly the most beautiful species of this family on this continent. A dull fish when young, it blossoms into a gorgeously-finned, colourful adult. A peaceful, shoaling fish, it takes pride of place in many home aquaria.

Catfish are the third largest family of freshwater fish in the world, and are, indeed, found almost world-wide, ranging from the tiny to the huge. Africa has a fair share, including the almost unbelievable upside-down catfish, which lives exactly as the name suggests. Swimming on its back, this fish is ideally suited to snatching aquatic insects from the surface of the water. Most fish are camouflaged, being darker around the upper regions, while lighter around the belly. The upside-down catfish (*Synodontis nigriventris*), as might be expected, neatly reverses this pattern.

Then there's the ropefish (*Erpetoichthys calabaricus*). Looks like a snake, or an eel, but is neither, gathering small worms and substrate-dwelling insect larvae.

But we must return to the cichlids for an ecological horror story, which will test the conscience of any would-be eco-warrior – the Nile perch.

These cichlids, which are lumped together under several genus names, and are by no means naturally confined to the Nile, have, in recent years, been artificially introduced into many other parts of Africa, and also South America and South-East Asia.

Tilapia, as they are generically known, are large fish – fast-growing and fast-breeding. It is also a popular food fish.

You may even have been introduced to one of these species in Britain. It now features on posh pub-grub menus where the offerings are a little more adventurous than the ubiquitous steak and ale pie, cod and chips, or the giant mixed grill.

These fish are, frankly, eating machines. Many of them show a distinct taste for vegetation – they have never been popular in home aquaria because of their habit of causing complete destruction in planted tanks – but they will greedily feast on any living creature small enough to be ingested.

In the wild, these tendencies cause the destruction of native fish populations, which are either eaten by the tilapia, or have their habitats destroyed. This has led to the near – and possibly actual – extinction of many local species.

The other side of the coin is this: For many indigenous people around the world, the introduction of tilapia has provided the only regular animal protein source available. The percentage of the population dependent on this fish in the Third World is impossible to calculate, but is certainly high.

Those who – perhaps rightly – deplore the ecological damage caused by the introduction of these fish, usually do so with three good meals a day inside them, and two cars parked in the garage. Those less fortunate may take a different view.

Congo tetra (*Phenacogrammus interruptus*)

Congo tetra (*Phenacogrammus interruptus*)

Jewel Cichlid (*Hemichromis bimaculatus*)

Kribensis (*Pelvicachromis pulcher*)

Maylandia lombardoi - an example of an east African cichlid

Ropefish or reed fish (*Erpetoichthys calabaricus*)

Upside Down Catfish (*Synodontis nigriventris*)

Tilapia spp.
All fish images courtesy:
Dr. Iggy Tavares

ACKNOWLEDEMENTS

Richard Freeman, Dr. Chris Clark, Chris Moiser, Lisa Dowley, Suzi Marsh, Oll Lewis (for going on the expedition), Corinna James (proofing and spell-checking), Mark North, Graham Inglis, Dr. Karl Shuker, Owen Burnham, Simon Wolstencroft, Nick Redfern, Kaye Phillips, Dave Phillips, Ross Phillips, Greg Phillips, Helen Bond, Jessica Bondall at Gambia Experience, the late John Downes, Rev R.J. Downes B.E.M., Assan Njie, Lamin Colly Dave Tanner, Dave Sutton, Fortean Times, John Gledson, Dan Garraway and Andy at CFZ TV.

PICTURE CREDITS

Dr. Karl Shuker, Lisa Dowley, Oll Lewis, Suzi Marsh, Chris Moiser, Mark North, Dr. Iggy Tavares, Ross Phillips, Greg Phillips, Jessica Bond.

Other books available from
CFZ PRESS

THE OWLMAN AND OTHERS - 30th Anniversary Edition
Jonathan Downes - ISBN 978-1-905723-02-7

£14.99

EASTER 1976 - Two young girls playing in the churchyard of Mawnan Old Church in southern Cornwall were frightened by what they described as a "nasty bird-man". A series of sightings that has continued to the present day. These grotesque and frightening episodes have fascinated researchers for three decades now, and one man has spent years collecting all the available evidence into a book. To mark the 30th anniversary of these sightings, Jonathan Downes, has published a special edition of his book.

DRAGONS - More than a myth?
Richard Freeman - ISBN 0-9512872-9-X

£14.99

First scientific look at dragons since 1884. It looks at dragon legends worldwide, and examines modern sightings of dragon-like creatures, as well as some of the more esoteric theories surrounding dragonkind. Dragons are discussed from a folkloric, historical and cryptozoological perspective, and Richard Freeman concludes that: "When your parents told you that dragons don't exist - they lied!"

MONSTER HUNTER
Jonathan Downes - ISBN 0-9512872-7-3

£14.99

Jonathan Downes' long-awaited autobiography, *Monster Hunter*... Written with refreshing candour, it is the extraordinary story of an extraordinary life, in which the author crosses paths with wizards, rock stars, terrorists, and a bewildering array of mythical and not so mythical monsters, and still just about manages to emerge with his sanity intact.......

MONSTER OF THE MERE
Jonathan Downes - ISBN 0-9512872-2-2

£12.50

It all starts on Valentine's Day 2002 when a Lancashire newspaper announces that "Something" has been attacking swans at a nature reserve in Lancashire. Eyewitnesses have reported that a giant unknown creature has been dragging fully grown swans beneath the water at Martin Mere. An intrepid team from the Exeter based Centre for Fortean Zoology, led by the author, make two trips – each of a week – to the lake and its surrounding marshlands. During their investigations they uncover a thrilling and complex web of historical fact and fancy, quasi Fortean occurrences, strange animals and even human sacrifice.

**CFZ PRESS, MYRTLE COTTAGE,
WOOLFARDISWORTHY BIDEFORD,
NORTH DEVON, EX39 5QR
w w w . c f z . o r g . u k**

Other books available from
CFZ PRESS

ONLY FOOLS AND GOATSUCKERS
Jonathan Downes - ISBN 0-9512872-3-0

£12.50

In January and February 1998 Jonathan Downes and Graham Inglis of the Centre for Fortean Zoology spent three and a half weeks in Puerto Rico, Mexico and Florida, accompanied by a film crew from UK Channel 4 TV. Their aim was to make a documentary about the terrifying chupacabra - a vampiric creature that exists somewhere in the grey area between folklore and reality. This remarkable book tells the gripping, sometimes scary, and often hilariously funny story of how the boys from the CFZ did their best to subvert the medium of contemporary TV documentary making and actually do their job.

WHILE THE CAT'S AWAY
Chris Moiser - ISBN: 0-9512872-1-4

£7.99

Over the past thirty years or so there have been numerous sightings of large exotic cats, including black leopards, pumas and lynx, in the South West of England. Former Rhodesian soldier Sam McCall moved to North Devon and became a farmer and pub owner when Rhodesia became Zimbabwe in 1980. Over the years despite many of his pub regulars having seen the "Beast of Exmoor" Sam wasn't at all sure that it existed. Then a series of happenings made him change his mind. Chris Moiser—a zoologist—is well known for his research into the mystery cats of the westcountry. This is his first novel.

THE BEAST AND I
Paul Crowther - ISBN 0-9512872-4-9

£7.99

In this extraordinarily funny book about the "Beast of Bodmin". Paul Crowther tells us of the adventures of "The Beast and I" and, along the way, introduces us to such engaging characters as the female vicar who risked life and limb to observe a beast at close quarters, the zoo-keeper with a portfolio of pictures of a domestic moggie, `Mr Angry` (who was annoyed to find a picture of his pet cat plastered all over the local newspaper, captioned as "The Beast"), and - of course - `Lara` the lynx of old London town.

BIG CATS IN BRITAIN YEARBOOK 2006
Edited by Mark Fraser - ISBN 978-1905723-01-0

£10.00

Big cats are said to roam the British Isles and Ireland even now as you are sitting and reading this. People from all walks of life encounter these mysterious felines on a daily basis in every nook and cranny of these two countries. Most are jet-black, some are white, some are brown, in fact big cats of every description and colour are seen by some unsuspecting person while on his or her daily business. 'Big Cats in Britain' are the largest and most active group in the British Isles and Ireland This is their first book. It contains a run-down of every known big cat sighting in the UK during 2005, together with essays by various luminaries of the British big cat research community which place the phenomenon into scientific, cultural, and historical perspective.

CFZ PRESS, MYRTLE COTTAGE, WOOLFARDISWORTHY BIDEFORD, NORTH DEVON, EX39 5QR
www.cfz.org.uk

Other books available from
CFZ PRESS

ANIMALS & MEN - Issues 1 - 5 - In the Beginning
Edited by Jonathan Downes - ISBN 0-9512872-6-5

£12.50

At the beginning of the 21st Century monsters still roam the remote, and sometimes not so remote, corners of our planet. It is our job to search for them. The Centre for Fortean Zoology [CFZ] is the only professional, scientific and full-time organisation in the world dedicated to cryptozoology - the study of unknown animals. Since 1992 the CFZ has carried out an unparalleled programme of research and investigation all over the world. We have carried out expeditions to Sumatra (2003 and 2004), Mongolia (2005), Puerto Rico (1998 and 2004), Mexico (1998), Thailand (2000), Florida (1998), Nevada (1999 and 2003), Texas (2003 and 2004), and Illinois (2004). An introductory essay by Jonathan Downes, notes putting each issue into a historical perspective, and a history of the CFZ.

THE BLACKDOWN MYSTERY
Jonathan Downes - ISBN 978-1-905723-00-3

£7.99

Intrepid members of the CFZ are up to the challenge, and manage to entangle themselves thoroughly in the bizarre trappings of this case. This is the soft underbelly of ufology, rife with unsavory characters, plenty of drugs and booze." That sums it up quite well, we think. A new edition of the classic 1999 book by legendary fortean author Jonathan Downes. In this remarkable book, Jon weaves a complex tale of conspiracy, anti-conspiracy, quasi-conspiracy and downright lies surrounding an air-crash and alleged UFO incident in Somerset during 1996. However the story is much stranger than that. This excellent and amusing book lifts the lid off much of contemporary forteana and explains far more than it initially promises.

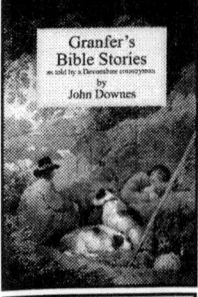

GRANFER'S BIBLE STORIES
John Downes - ISBN 0-9512872-8-1

£7.99

Bible stories in the Devonshire vernacular, each story being told by an old Devon Grandfather - 'Granfer'. These stories are now collected together in a remarkable book presenting selected parts of the Bible as one more-or-less continuous tale in short 'bite sized' stories intended for dipping into or even for bed-time reading. `Granfer` treats the biblical characters as if they were simple country folk living in the next village. Many of the stories are treated with a degree of bucolic humour and kindly irreverence, which not only gives the reader an opportunity to re-evaluate familiar tales in a new light, but do so in both an entertaining and a spiritually uplifting manner.

FRAGRANT HARBOURS DISTANT RIVERS
John Downes - ISBN 0-9512872-5-7

£12.50

Many excellent books have been written about Africa during the second half of the 19th Century, but this one is unique in that it presents the stories of a dozen different people, whose interlinked lives and achievements have as many nuances as any contemporary soap opera. It explains how the events in China and Hong Kong which surrounded the Opium Wars, intimately effected the events in Africa which take up the majority of this book. The author served in the Colonial Service in Nigeria and Hong Kong, during which he found himself following in the footsteps of one of the main characters in this book; Frederick Lugard – the architect of modern Nigeria.

**CFZ PRESS, MYRTLE COTTAGE,
WOOLFARDISWORTHY BIDEFORD,
NORTH DEVON, EX39 5QR
w w w . c f z . o r g . u k**

THE CENTRE FOR FORTEAN ZOOLOGY

The Centre for Fortean Zoology is the world's only professional and scientific organisation dedicated to research into unknown animals. Although we work all over the world, we carry out regular work in the United Kingdom and abroad, investigating accounts of strange creatures.

THAILAND 2000
An expedition to investigate the legendary creature known as the Naga

SUMATRA 2003
'Project Kerinci'
In search of the bipedal ape Orang Pendek

MONGOLIA 2005
'Operation Death Worm'
An expedition to track the fabled 'Allghoi Khorkhoi' or Death Worm

Led by scientists, the CFZ is staffed by volunteers and is always looking for new members.

To apply for a <u>FREE</u> information pack about the organisation and details of how to join and receive our quarterly journal, plus information on current and future projects, expeditions and events.

Send a stamp addressed envelope to:

THE CENTRE FOR FORTEAN ZOOLOGY
MYRTLE COTTAGE, WOOLSERY,
BIDEFORD, NORTH DEVON EX39 5QR.

or alternatively visit our website at: w w w . c f z . o r g . u k

www.ingramcontent.com/pod-product-compliance
Lightning Source LLC
Chambersburg PA
CBHW062154080426
42734CB00010B/1682